HRW
ALGEBRA ONE
INTERACTIONS
COURSE 1
PRACTICE WORKBOOK

HOLT, RINEHART AND WINSTON
Harcourt Brace & Company

Austin • New York • Orlando • Atlanta • San Francisco • Boston • Dallas • Toronto • London

To the Student

HRW Algebra One Interactions Course 1 Practice Workbook is designed to provide additional practice of the skills taught in each lesson of your textbook. On each page you will practice the skills from one particular lesson. There are approximately 10 to 35 practice items on each page. These items include practice of both the basic skills and the mathematical applications taught in the lesson.

Printed in the United States of America

ISBN 0-03-051258-1

3   4   5   6   7   066   00   99   98   97

TABLE OF CONTENTS

NAME _____ CLASS _____ DATE _____

 Practice
1.1 Representing Number Patterns

1. There are seven people in a room. Each person shakes hands with each of the other people in the room. How many handshakes take place? _____

2. There are six players in a backgammon tournament. If each player must play every other player, how many games need to be played? _____

3. Find the sum $1 + 2 + 3 + 4 + 5 + 6 + 7 + 8$ using the geometric dot pattern. _____

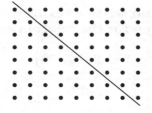

4. Find the sum $1 + 2 + 3 + 4 + 5 + 6 + 7 + 8 + 9 + 10$ using the geometric dot pattern. _____

Find the next three terms in each sequence. Then explain the pattern used to find the terms for each.

5. 12, 17, 22, 27, 32, _____, _____, _____ _____

6. 2, 4, 8, 16, 32, _____, _____, _____ _____

7. 100, 87, 74, 61, 48, _____, _____, _____ _____

8. 2, 2, 4, 6, 10, 16, _____, _____, _____ _____

9. Complete the table.

Number	2	4	6	8	10	12	14	16	18
Sum	2	6	12	20					

10. Examine the pattern in the table for Exercise 9. Guess the sum of the first 100 even numbers. Explain your method.

_____

 **Practice**

1.2 Exploring Problems With Tables and Equations

Write an algebraic expression for each verbal expression.

1. Each ticket costs $5.   _____

2. Julie makes $8 per hour.   _____

3. Each box holds 12 cans. _____

4. Art rode 6 miles each hour. _____

Complete the table, and write an expression for the process.
Then write an equation for the relationship.

5. Mark rents a bicycle for $10 plus $2.50 per hour.

Number of Hours h	Process _____	Cost c
1	10 + 2.5(1)	
2	10 + 2.5(_____)	
3	10 + 2.5(_____)	
4	10 + 2.5(_____)	
5	10 + 2.5(_____)	
10	10 + 2.5(_____)	

6. Write an equation for the cost using the variables c and h. $c =$ _____

Build a table of values by substituting 1, 2, 3, 4, 5, and 10 for x.

7. $y = 4x$

x	1	2	3	4	5	10
y						

8. $y = x + 5$

x	1	2	3	4	5	10
y						

9. $y = 3x - 3$

x	1	2	3	4	5	10
y						

10. $y = 2x + 1$

x	1	2	3	4	5	10
y						

Practice
1.3 Solving Problems Using Tables and Charts

Complete the table of values by substituting 10, 20, 30, and 40 for the variable in the expression.

1. $3.1x$

x	Process	Value
10		
20		
30		
40		

2. $6c - 5$

c	Process	Value
10		
20		
30		
40		

The cost of renting a computer is given by the formula $c = 100 + 99m$, where c is the total cost and m is the number of months that the computer is rented.

3. Complete the table.

Number of months, m	Process $100 + 99m$	Cost, c
1	$100 + 99(1)$	
2	$100 + 99(\underline{\hspace{1.5cm}})$	
3		
4		
5		

4. Suppose you can afford to spend $397 renting the computer. What equation describes this situation, and how many months can you rent

the computer? _____

5. Make a bar chart with the data from your table.

6. Use the bar chart from Exercise 7 to determine how many months you could rent the computer if you had $500.

_____

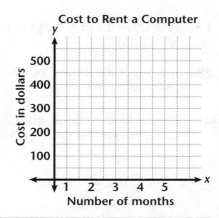

Cost to Rent a Computer

Practice
1.4 Exploring Variables and Equations

Use guess-and-check to solve each equation.

1. $6x + 11 = 47$ _____

2. $23 = 11x + 1$ _____

3. $52 = 4x - 4$ _____

4. $48x + 6 = 246$ _____

Identify each as an expression or a variable.

5. $2x + 1$ _____

6. r _____

7. q _____

8. $4v$ _____

If T-shirts cost \$5.50 each, find the cost of

9. 2 T-shirts. _____

10. 7 T-shirts. _____

11. 15 T-shirts. _____

12. t T-shirts. _____

Complete each table to show the substitutions of 1, 2, 3, 4, and 5 for the variables in the expressions.

13. $7y$

y					
$7y$					

14. $m + 6$

m					
$m + 6$					

15. $5y + 1$

y					
$5y + 1$					

16. $6t - 1$

t					
$6t - 1$					

17. $3r + 2$

r					
$3r + 2$					

18. $4.5c$

c					
$4.5c$					

19. If tickets for a play cost \$15 each, how many tickets can you buy with \$120? _____

20. If tickets for a movie cost \$6.50 each, how many tickets can you buy with \$59? _____

21. If tickets for each ride at an amusement park cost \$2.25 each, how many tickets can you buy with \$20? _____

Practice
1.5 Exploring Factors and Divisibility Patterns

Write the list of factors for each number. Circle the prime factors.

1. 42 _____   **2.** 24 _____

3. 18 _____   **4.** 81 _____

5. 25 _____   **6.** 28 _____

Determine whether the number is divisible by 2, by 5, and by 10.

7. 445 _____   **8.** 334 _____   **9.** 6922 _____

10. 846 _____   **11.** 930 _____   **12.** 2398 _____

13. 1870 _____   **14.** 5135 _____   **15.** 12,444 _____

Determine whether the number is divisible by 2, by 3, and by 6.

16. 1568 _____   **17.** 2640 _____   **18.** 28,002 _____

19. 15,087 _____   **20.** 70,002 _____   **21.** 9778 _____

22. 1215 _____   **23.** 16,978 _____   **24.** 68,342 _____

Determine whether the number is divisible by 3, by 6, and by 9.

25. 990 _____   **26.** 1974 _____   **27.** 15,330 _____

28. 855 _____   **29.** 10,269 _____   **30.** 48,642 _____

Determine whether each number is divisible by 15.

31. 150 _____   **32.** 955 _____   **33.** 1060 _____

34. 660 _____   **35.** 465 _____   **36.** 980 _____

Determine whether each number is prime, composite, or neither.

37. 11 _____   **38.** 2 _____   **39.** 195 _____

40. 61 _____   **41.** 153 _____   **42.** 98 _____

43. 1 _____   **44.** 87 _____   **45.** 63 _____

 **Practice**

1.6 Exponents and Prime Factorization

Use exponents to rewrite each expression.

1. $6 \times 6 \times 6 \times 6 \times 6$ _____

2. $8 \times 8 \times 8$ _____

3. $3 \cdot 3 \cdot 3 \cdot 3 \cdot 5 \cdot 5$ _____

4. $2 \cdot 2 \cdot 7 \cdot 7 \cdot 7 \cdot 11 \cdot 11$ _____

5. $a \cdot a \cdot a \cdot a \cdot c \cdot c$ _____

6. $z \cdot z \cdot z + 4$ _____

Evaluate each power.

7. 3^4 _____

8. 6^3 _____

9. 2^7 _____

10. 10^5 _____

11. 8^3 _____

12. 7^2 _____

Use the formula $V = e^3$ to find the volume of a cube with the following edge lengths.

13. 5 inches _____

14. 8 centimeters _____

15. 4 feet _____

16. 1.1 miles _____

17. 4.3 meters _____

18. 2.2 yards _____

Write the prime factorization of each number.

19. 24 _____

20. 81 _____

21. 150 _____

22. 90 _____

23. 54 _____

24. 144 _____

25. 102 _____

26. 42 _____

Evaluate each expression if $a = 3$, $b = 4$, and $c = 6$.

27. abc _____

28. ab^2c _____

29. abc^2 _____

30. a^3b^2c _____

31. $a^2b^2c^2$ _____

32. ab^3c^2 _____

Use your calculator to find the prime factorization of each number. Write your answers in exponential form.

33. 5100 _____

34. 18,660 _____

35. 27,027 _____

36. 91,200 _____

Practice
1.7 Order of Operations

Place inclusion symbols according to the correct order of operations in order to make each equation true. Tell how you would use a calculator to check your answer.

1. $27 + 5 \cdot 8 - 6 = 37$ _____

2. $12 \cdot 1 + 5 \div 12 = 6$ _____

3. $4 \cdot 5 - 3 + 2 = 10$ _____

4. $3 \cdot 4 + 2 \div 6 = 3$ _____

Simplify by using the correct order of operations. If answers are not exact, round them to three decimal places.

5. $16 + 4 \cdot 8 + 2$ _____

6. $16 + 4 \cdot (8 + 2)$ _____

7. $(16 + 4) \cdot (8 + 2)$ _____

8. $(16 + 4) \cdot 8 + 2$ _____

9. $63 \cdot 30 + 43$ _____

10. $91(2.5) + 3.8$ _____

11. $0.2(2.5) + 8$ _____

12. $16 \cdot 37 + 88 \cdot 49$ _____

13. $5.6(3.6) + 8.9(3.7)$ _____

14. $8.5(3.4 + 2.6)$ _____

15. $16 + 8 \div 2$ _____

16. $4 \cdot 6 \div 12 + 10$ _____

17. $\dfrac{12 + 6}{4 + 2}$ _____

18. $\dfrac{5 + 3 \cdot 5}{5}$ _____

19. $12 + 6 \div 4 + 2$ _____

20. $89 \div 3 + 5$ _____

21. $36 - 6 \cdot 3 \div 18 \div 3$ _____

22. $9 - 3 \div 4 + 2 \cdot 12 + 6 \div 2 \cdot 3$ _____

23. $4 + 1 \cdot 4^2 - 3$ _____

24. $6 + 3^3 - 18 \div 7$ _____

Given that a is 3, b is 6, and c is 5, evaluate each expression.

25. $b - a + c$ _____

26. $a + b - c$ _____

27. $a \cdot b + a \cdot c$ _____

28. $b \div a + a \cdot c$ _____

29. $a^2 + c^2$ _____

30. $b^2 - a^2$ _____

31. $(a + b) \div c$ _____

32. $a + b^2 - c$ _____

33. $a^2 + (b^2 - c)$ _____

34. $c^2 - (b - a)$ _____

NAME _____ CLASS _____ DATE _____

 **Practice**
1.8 Exploring Properties and Mental Computation

Complete each step, and name the property used.

1. $(24 + 68) + 66$

= $(68 + \underline{\hspace{1.5cm}}) + 66$ Commutative Property

= $68 + (24 + \underline{\hspace{1.5cm}})$ $\underline{\hspace{2cm}}$ Property

= $68 + \underline{\hspace{1.5cm}}$

= $\underline{\hspace{1.5cm}}$

2. $35(3 + 5)$

= $35 \cdot \underline{\hspace{1.5cm}} + \underline{\hspace{1.5cm}} \cdot 5$ $\underline{\hspace{2cm}}$ Property

= $\underline{\hspace{1.5cm}} + \underline{\hspace{1.5cm}}$

= $\underline{\hspace{1.5cm}}$

Use mental math to find each sum or product. Show your work and explain each step.

3. $(46 + 28) + 24$ $\underline{\hspace{3cm}}$ **4.** $(96 \cdot 4) \cdot 5$ $\underline{\hspace{3cm}}$

5. $(828 + 386) + 412$ $\underline{\hspace{3cm}}$ **6.** $2 \cdot (137 \cdot 5)$ $\underline{\hspace{3cm}}$

Name the property illustrated.

7. $46 + 12 = 12 + 46$ $\underline{\hspace{5cm}}$

8. $23 + (17 + 34) = (23 + 17) + 34$ $\underline{\hspace{5cm}}$

9. $4(2.3 + 4.9) = 4(2.3) + 4(4.9)$ $\underline{\hspace{5cm}}$

10. $6(3x) = (6 \cdot 3)x$ $\underline{\hspace{5cm}}$

11. $5 \cdot (12 \cdot 4) = 5 \cdot (4 \cdot 12)$ $\underline{\hspace{5cm}}$

12. $6 \cdot 300 + 6 \cdot 80 = 6(300 + 80)$ $\underline{\hspace{5cm}}$

8   **Practice**   **HRW Algebra One Interactions Course 1**

Practice
2.1 Integers and the Number Line

Determine whether each number is an integer.

1. 4 _____

2. −1 _____

3. 2.3 _____

4. $-\frac{3}{4}$ _____

5. 0 _____

6. −25 _____

Write an integer to represent each amount.

7. 32 degrees above zero _____

8. a deposit of $50 _____

9. a loss of 2 yards _____

10. a rise in temperature of 10 degrees _____

11. 6 years from now _____

12. a withdrawal of $20 _____

13. a price decrease of $15 _____

14. 9 years ago _____

15. a weight gain of 5 pounds _____

16. a depth of 3 feet _____

Write the opposite of each number.

17. 15 _____

18. −4 _____

19. −1 _____

20. 47 _____

21. −59 _____

22. 13 _____

Find the absolute value.

23. $|-7|$ _____

24. $|6|$ _____

25. $|-3|$ _____

26. $|-36|$ _____

27. $|1|$ _____

28. $|97|$ _____

29. $|0|$ _____

30. $|-127|$ _____

31. $|500|$ _____

Write the indicated change in temperature as a positive or negative integer.

32. starts at −5 degrees and ends at −10 degrees _____

33. starts at −9 degrees and ends at −2 degrees _____

34. starts at −7 degrees and ends at −1 degree _____

35. starts at −4 degrees and ends at −8 degrees _____

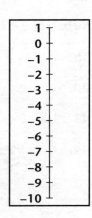

 **Practice**

2.2 Exploring Integer Addition

Use algebra tiles to find each sum.

1. $-3 + 2$ _____

2. $-4 + (-1)$ _____

3. $4 + (-5)$ _____

4. $-6 + 6$ _____

5. $-3 + (-7)$ _____

6. $9 + (-4)$ _____

7. $7 + (-4)$ _____

8. $-7 + 9$ _____

9. $-8 + 3$ _____

10. $10 + (-9)$ _____

11. $-1 + (-9)$ _____

12. $(-6) + (-2)$ _____

13. $-4 + (-1) + 3$ _____

14. $5 + (-1) + (-2)$ _____

15. $-3 + (-4) + 2$ _____

Find each sum.

16. $-35 + 40$ _____

17. $15 + (-28)$ _____

18. $60 + (-18)$ _____

19. $-17 + (-19)$ _____

20. $42 + (-56)$ _____

21. $-34 + 28$ _____

22. $59 + (-59)$ _____

23. $-86 + 85$ _____

24. $-45 + (-45)$ _____

25. $-68 + (-15)$ _____

26. $-3 + (-1) + (-2)$ _____

27. $4 + (-7) + (-4)$ _____

28. $-54 + 63 + (-20)$ _____

29. $-78 + (-78) + 50$ _____

30. $-6 + (-42) + 24$ _____

31. $-24 + (-62) + (-11)$ _____

32. $-5 + |-4|$ _____

33. $|5| + |-4|$ _____

34. $|-5| + |-4|$ _____

35. $|-5| + |4| + (-9)$ _____

36. $-48 + |-64| + (-32)$ _____

37. $-568 + (-43) + |-57|$ _____

Substitue 4 for *a*, −6 for *b*, and 3 for *c*. Evaluate each expression.

38. $a + (b + c)$ _____

39. $a + |b + c|$ _____

40. $(a + c) + b$ _____

41. $(a + c) + |b|$ _____

42. $|a + b| + c$ _____

43. $|a + c| + b$ _____

44. $|c| + b + a$ _____

45. $a + |b| + |c|$ _____

Practice
2.3 Solving Equations and Comparing Integers

Use guess-and-check to solve each equation.

1. $x + 4 = 3$ _____ **2.** $x + 7 = -1$ _____ **3.** $6 + x = 3$ _____

4. $3 + x = -3$ _____ **5.** $-2 + x = -5$ _____ **6.** $x + (-4) = 9$ _____

7. $x + 8 = 0$ _____ **8.** $-4 + x = 6$ _____ **9.** $-1 + x = -6$ _____

**Write two inequalities for each pair of integers. Use both the <
and > symbols.**

10. $6, -6$ _____ **11.** $-3, 0$ _____

12. $-3, -8$ _____ **13.** $-4, 1$ _____

14. $7, -7$ _____ **15.** $-9, -5$ _____

16. $0, -4$ _____ **17.** $3, -6$ _____

**Show each integer on the number line provided. Then list the
integers in ascending order (from least to greatest).**

18. $6, -6, 2, -1$

_____

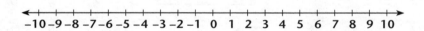

19. $-9, 3, 0, -2, 6, -7$

_____

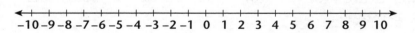

20. $5, 0, -4, 4, -5$

_____

Use guess-and-check to solve each equation.

21. $-10 + x = -1$ _____ **22.** $v + 14 = 9$ _____ **23.** $45 - x = -5$ _____

24. $37 + z = -32$ _____ **25.** $19 - r = -24$ _____ **26.** $-27 + x = -6$ _____

27. $2a = -6$ _____ **28.** $3a = -15$ _____ **29.** $6y = -24$ _____

30. $36 \div t = -9$ _____ **31.** $-14 \div w = -7$ _____ **32.** $3a + 5 = -1$ _____

33. $2c + 1 = -9$ _____ **34.** $2c - 1 = -9$ _____ **35.** $5y - 10 = 0$ _____

Practice
2.4 Exploring Integer Subtraction

Use algebra tiles to find each difference.

1. $3 - 2$ _____

2. $-3 - (-2)$ _____

3. $3 - (-2)$ _____

4. $-3 - 2$ _____

5. $-4 - 3$ _____

6. $-8 - (-2)$ _____

7. $4 - 7$ _____

8. $6 - (-5)$ _____

9. $-8 - (-8)$ _____

Evaluate each expression.

10. $56 - 2$ _____

11. $53 - (-8)$ _____

12. $26 - (-26)$ _____

13. $-85 - (-34)$ _____

14. $-64 - 73$ _____

15. $-56 + (-42)$ _____

16. $58 - (-58)$ _____

17. $-49 - 18$ _____

18. $-24 + 47 + (-24)$ _____

19. $-13 + 19 - (-25)$ _____

20. $-66 - 66 + 6$ _____

21. $86 - (-15) - 9$ _____

22. $45 - (-27) - (-17)$ _____

23. $-29 - 16 - (-37)$ _____

24. $72 - 56 - 13$ _____

25. $-99 + 16 - (-24)$ _____

Substitute 4 for x, -4 for y, and -12 for z. Evaluate each expression.

26. $z - y$ _____

27. $x + z$ _____

28. $x + y - z$ _____

29. $(x - y) + z$ _____

30. $y + z$ _____

31. $(y + z) - x$ _____

32. $y - z$ _____

33. $(x - z) - y$ _____

34. $z - z - z$ _____

35. $(x + y) + (y - z)$ _____

36. $y - (x - z)$ _____

37. $x - x - x - x$ _____

Find the distance between each pair of points on the number line.

38. $6, 10$ _____

39. $-5, 2$ _____

40. $-35, -38$ _____

41. $-13, 26$ _____

42. $-44, -29$ _____

43. $-15, 73$ _____

Practice

2.5 Exploring Integer Multiplication and Division

Evaluate.

1. $(4)(-5)$ _____

2. $(3)(-5)$ _____

3. $(2)(-5)$ _____

4. $(-1)(-5)$ _____

5. $(-2)(-5)$ _____

6. $(-3)(-5)$ _____

7. $(4)(-3)$ _____

8. $(3)(-3)$ _____

9. $(2)(-3)$ _____

10. $(-1)(-3)$ _____

11. $(-2)(-3)$ _____

12. $(-3)(-3)$ _____

13. $(-11)(-4)$ _____

14. $(-11) - (-4)$ _____

15. $(-7) - (-4)$ _____

16. $(-42) \div (-3)$ _____

17. $(-35)(22)$ _____

18. $(-27)(-1.3)$ _____

19. $(-240) \div (-8)$ _____

20. $(-240) + (-8)$ _____

21. $(-0.5)(-12)$ _____

22. $(-2.1) \div (-7)$ _____

23. $(6)(5)(-7)$ _____

24. $(-3)[(-1) + (-5)]$ _____

25. $(-8) \div [5 + (-3)]$ _____

26. $(-2.5) \div (-4)$ _____

27. $(-7)(-3)(6)$ _____

28. $(-4488) \div (136)$ _____

29. $(-5)(5)(5) \div (5)$ _____

30. $(-2)[5 + (-5)]$ _____

31. $\dfrac{(8)(-1)}{-8}$ _____

32. $\dfrac{(-2)(-14)}{7}$ _____

33. $\dfrac{(-2)(20)(-40)}{-10}$ _____

Tell whether each statement is true or false.

34. The product of two negatives is positive. _____

35. The quotient of two negatives is positive. _____

36. The average of a set of negative numbers is positive. _____

37. The difference of two positives is always positive. _____

38. The sum of two positives is positive. _____

Stephanie opened a savings account with a $35 deposit. She made a total of 6 additional deposits of $15 each and withdrawals of $5, $10, and $15.

39. What is the total amount that Stephanie deposited in her account after her initial deposit? _____

40. What is the total amount that Stephanie withdrew from her account after her initial deposit? _____

41. What is the total amount currently in Stephanie's account? _____

Practice

2.6 Solving Problems With Equations and Graphs

Make a table of values and ordered pairs for each equation by substituting integer values from −3 to 3 for x.

1. $y = 4 - x$

x							
y							

2. $y = 2x - 4$

x							
y							

3. $y = -3x + 7$

x							
y							

Write the ordered pair that represents each point.

4. A _____

5. B _____

6. C _____

7. D _____

8. E _____

9. F _____

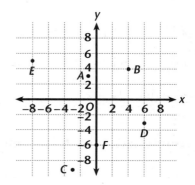

Graph each equation by using integer values of x from −2 to 3. Use your graph to find the value of y when x is 4.

10. $y = 3 - x$

x						
y						

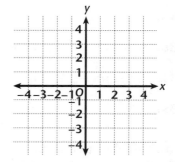

Write a linear equation to describe each set of data.

11.

x	−3	−2	−1	0	1	2	3
y	15	10	5	0	−5	−10	−15

_____

12.

x	−2	−1	0	1	2	3
y	−3	−2	−1	0	1	2

_____

Practice
2.7 Using Differences to Predict and Generalize

Find the next two terms of each sequence.

1. 32, 29, 26, 23, 20, _____, _____

2. 500, 483, 490, 473, 480, _____, _____

3. 13, 52, 91, 130, 169, _____, _____

4. 89, 74, 59, 44, _____, _____

5. 15, 17, 14, 16, 13, _____, _____

6. 29, 24, 30, 25, 31, _____, _____

7. 85, 75, 73, 63, 61, _____, _____

8. 6, 9, 14, 21, 30, _____, _____

9. 999, 949, 899, 849, _____, _____

10. 125, 250, 375, 500, _____, _____

11. 20, 29, 41, 56, 74, _____, _____

12. 17, 19, 22, 26, 31, _____, _____

13. 5, 12, 20, 29, 39, _____, _____

14. The first three terms of a sequence are 4, 7, and 12. Find the first and second differences. Assuming the second differences are constant, what are the next three terms? _____

15. The first three terms of a sequence are 1, 7, and 18. Find the first and second differences. Assuming the second differences are constant, what are the next three terms? _____

16. The first three terms of a sequence are 6, 10, and 17. The second differences are a constant 3. What are the next three terms of the sequence? _____

17. The first three terms of a sequence are 3, 5, and 11. The second differences are a constant 4. What are the next three terms of the sequence? _____

Practice
3.1 Introduction to Rational Numbers

Use fraction bars to tell whether each fraction is closest to 0, $\frac{1}{2}$, or 1.

1. $\frac{9}{10}$ _____

2. $\frac{1}{3}$ _____

3. $\frac{3}{5}$ _____

4. $\frac{3}{8}$ _____

5. $\frac{5}{6}$ _____

6. $\frac{1}{10}$ _____

Use fraction bars to compare the fractions, using the symbol <, >, or =.

7. $\frac{1}{2}$ and $\frac{3}{4}$ _____

8. $\frac{1}{4}$ and $\frac{1}{5}$ _____

9. $\frac{2}{3}$ and $\frac{5}{6}$ _____

10. $\frac{3}{8}$ and $\frac{3}{10}$ _____

11. $\frac{6}{12}$ and $\frac{1}{2}$ _____

12. $\frac{9}{10}$ and $\frac{4}{5}$ _____

13. $\frac{6}{8}$ and $\frac{2}{3}$ _____

14. $\frac{4}{12}$ and $\frac{2}{3}$ _____

15. $\frac{6}{8}$ and $\frac{3}{5}$ _____

16. $\frac{5}{10}$ and $\frac{4}{8}$ _____

17. $\frac{3}{5}$ and $\frac{6}{12}$ _____

18. $\frac{2}{3}$ and $\frac{3}{5}$ _____

19. $\frac{3}{5}$ and $\frac{7}{10}$ _____

20. $\frac{1}{3}$ and $\frac{1}{5}$ _____

21. $\frac{5}{12}$ and $\frac{1}{2}$ _____

Tell whether each fraction is closest to $\frac{1}{4}$, $\frac{1}{2}$, or $\frac{3}{4}$.

22. $\frac{5}{6}$ _____

23. $\frac{5}{12}$ _____

24. $\frac{3}{5}$ _____

25. $\frac{7}{12}$ _____

26. $\frac{11}{12}$ _____

27. $\frac{6}{10}$ _____

28. $\frac{4}{5}$ _____

29. $\frac{3}{10}$ _____

30. $\frac{1}{8}$ _____

31. $\frac{7}{10}$ _____

32. $\frac{6}{8}$ _____

33. $\frac{1}{6}$ _____

Use fraction bars to compare each fraction to $\frac{1}{2}$, using the symbol <, >, or =.

34. $\frac{1}{3}$ _____

35. $\frac{4}{5}$ _____

36. $\frac{10}{11}$ _____

37. $\frac{3}{8}$ _____

38. $\frac{6}{12}$ _____

39. $\frac{3}{10}$ _____

40. $\frac{9}{10}$ _____

41. $\frac{3}{6}$ _____

42. $\frac{8}{12}$ _____

43. $\frac{2}{5}$ _____

44. $\frac{3}{4}$ _____

45. $\frac{7}{10}$ _____

Practice
3.2 Using Equivalent Fractions

Write each fraction in lowest terms.

1. $\frac{4}{6}$ _____ **2.** $\frac{2}{8}$ _____ **3.** $\frac{3}{9}$ _____ **4.** $\frac{4}{16}$ _____

5. $\frac{10}{12}$ _____ **6.** $\frac{18}{20}$ _____ **7.** $\frac{9}{15}$ _____ **8.** $\frac{15}{35}$ _____

9. $\frac{100}{150}$ _____ **10.** $\frac{36}{48}$ _____ **11.** $\frac{81}{90}$ _____ **12.** $\frac{64}{72}$ _____

13. $\frac{144}{196}$ _____ **14.** $\frac{48}{56}$ _____ **15.** $\frac{27}{45}$ _____ **16.** $\frac{16}{64}$ _____

17. $\frac{300}{400}$ _____ **18.** $\frac{49}{63}$ _____ **19.** $\frac{125}{175}$ _____ **20.** $\frac{72}{81}$ _____

21. $\frac{50}{175}$ _____ **22.** $\frac{22}{121}$ _____ **23.** $\frac{39}{169}$ _____ **24.** $\frac{56}{70}$ _____

Use the LCD to compare each pair of fractions, using the symbol $<$, $>$, or $=$.

25. $\frac{2}{3}$ and $\frac{1}{6}$ _____ **26.** $\frac{2}{7}$ and $\frac{3}{14}$ _____ **27.** $\frac{1}{2}$ and $\frac{5}{12}$ _____

28. $\frac{3}{5}$ and $\frac{7}{15}$ _____ **29.** $\frac{30}{36}$ and $\frac{5}{6}$ _____ **30.** $\frac{3}{4}$ and $\frac{9}{16}$ _____

31. $\frac{7}{10}$ and $\frac{4}{5}$ _____ **32.** $\frac{7}{15}$ and $\frac{3}{10}$ _____ **33.** $\frac{2}{14}$ and $\frac{1}{5}$ _____

34. $\frac{2}{3}$ and $\frac{4}{7}$ _____ **35.** $\frac{12}{16}$ and $\frac{3}{4}$ _____ **36.** $\frac{3}{8}$ and $\frac{2}{3}$ _____

Write each list of fractions in order from least to greatest.

37. $\frac{3}{4}, \frac{2}{5}, \frac{4}{5}, \frac{1}{2}, \frac{1}{4},$ and $\frac{3}{10}$ _____

38. $-\frac{3}{8}, -\frac{1}{3}, \frac{1}{2}, -\frac{1}{4}, \frac{3}{4},$ and $-\frac{2}{3}$ _____

39. $\frac{3}{16}, \frac{3}{4}, \frac{1}{8}, \frac{1}{2}, \frac{9}{16},$ and $\frac{5}{8}$ _____

40. $-\frac{5}{6}, \frac{5}{12}, -\frac{1}{2}, \frac{5}{6}, -\frac{3}{4},$ and $\frac{1}{3}$ _____

41. $\frac{1}{2}, \frac{1}{9}, \frac{1}{5}, \frac{1}{3}, \frac{1}{4}, \frac{1}{10},$ and $\frac{1}{25}$ _____

42. What happens to the size of each fraction in Exercise 41 as the denominator decreases?

_____

Practice
3.3 Exploring Decimals

Write each decimal as a fraction or mixed number in simplest form.

1. 0.19 _____ **2.** 0.7 _____ **3.** 0.21 _____

4. 0.8 _____ **5.** 0.35 _____ **6.** 2.3 _____

7. 6.9 _____ **8.** 0.123 _____ **9.** 10.9 _____

10. 15.74 _____ **11.** 0.475 _____ **12.** 130.8 _____

13. 4.017 _____ **14.** 250.88 _____ **15.** 36.22 _____

16. 367.42 _____ **17.** 56.56 _____ **18.** 0.006 _____

19. 0.0058 _____ **20.** 2.0009 _____ **21.** 16.0198 _____

Write each fraction as a terminating or repeating decimal.

22. $\frac{4}{5}$ _____ **23.** $\frac{7}{12}$ _____ **24.** $-\frac{5}{6}$ _____

25. $-\frac{1}{4}$ _____ **26.** $\frac{9}{11}$ _____ **27.** $-\frac{8}{9}$ _____

28. $3\frac{1}{3}$ _____ **29.** $-1\frac{3}{7}$ _____ **30.** $-10\frac{3}{4}$ _____

Order the decimals from least to greatest.

31. 0.5 0.46 0.42 0.425 0.55 _____

32. −0.076 −0.76 −0.6 −0.06 −0.006 _____

33. 7.201 7.21 7.021 7.1 7.01 _____

34. −24.3 −24 −24.03 −23.4 −23.04 _____

Order the numbers from greatest to least.

35. 6.8 $6\frac{3}{4}$ 6.08 $6\frac{2}{3}$ $6\frac{5}{6}$ _____

36. 0.59 $-\frac{1}{2}$ −0.9 $\frac{3}{5}$ $-\frac{4}{7}$ $-\frac{3}{4}$ _____

37. $\frac{12}{7}$ 1.7 1.177 $1\frac{3}{4}$ $1\frac{2}{3}$ _____

Practice

3.4 Exploring Addition and Subtraction of Rational Numbers

Estimate the perimeter of each picture frame with the indicated dimensions. Then compute the exact perimeter.

1. $7\frac{1}{2}$ by $8\frac{3}{4}$ _____

2. $6\frac{5}{8}$ by $10\frac{1}{2}$ _____

3. $10\frac{3}{8}$ by $15\frac{1}{4}$ _____

4. $24\frac{5}{16}$ by $18\frac{1}{4}$ _____

Find two mixed numbers that have the given sum.

5. $2\frac{5}{6}$ _____

6. $6\frac{1}{3}$ _____

7. $\frac{3}{4}$ _____

8. $3\frac{2}{5}$ _____

9. $-4\frac{3}{7}$ _____

10. $-8\frac{1}{2}$ _____

11. $\frac{6}{7}$ _____

12. $-\frac{11}{12}$ _____

13. $3\frac{5}{9}$ _____

Find two mixed numbers that have the given difference.

14. $2\frac{5}{6}$ _____

15. $6\frac{1}{3}$ _____

16. $\frac{3}{4}$ _____

17. $3\frac{2}{5}$ _____

18. $-4\frac{3}{7}$ _____

19. $-8\frac{1}{2}$ _____

20. $\frac{6}{7}$ _____

21. $-\frac{11}{12}$ _____

22. $3\frac{5}{9}$ _____

Add or subtract. Check each answer by using a calculator.

23. $\frac{2}{3} + \frac{1}{2}$ _____

24. $\frac{7}{8} - \frac{1}{4}$ _____

25. $\frac{4}{5} + \frac{2}{3}$ _____

26. $\frac{9}{10} + \frac{3}{5}$ _____

27. $\frac{1}{4} + \frac{2}{15}$ _____

28. $6 - \frac{3}{8}$ _____

29. $\frac{6}{7} - \frac{2}{5}$ _____

30. $4\frac{1}{2} - \frac{2}{3}$ _____

31. $6\frac{3}{4} + 1\frac{2}{3}$ _____

32. $\frac{2}{5} + \frac{6}{7} - \frac{1}{2}$ _____

33. $\frac{5}{9} - \frac{1}{3} + 1\frac{1}{6}$ _____

34. $9 - 2\frac{7}{15}$ _____

35. $-\frac{1}{9} - \frac{5}{6}$ _____

36. $-\frac{3}{7} + 2\frac{1}{2}$ _____

37. $-5\frac{4}{5} + \frac{7}{15}$ _____

38. $8\frac{1}{5} - 3\frac{7}{10}$ _____

39. $-\frac{9}{11} + 4\frac{2}{3}$ _____

40. $\frac{7}{8} - 8\frac{1}{2} + \frac{5}{16}$ _____

41. $-\frac{2}{3} + \left(-2\frac{3}{8}\right) + \left(-\frac{1}{2}\right)$ _____

42. $\frac{2}{9} + \left(-3\frac{5}{12}\right) + \left(-2\frac{1}{4}\right)$ _____

Practice
3.5 Multiplying and Dividing Rational Numbers

Estimate each product. Then find the actual product.

1. $8 \cdot \frac{2}{3}$ _____

2. $\frac{8}{9} \cdot 3$ _____

3. $7 \cdot \frac{4}{5}$ _____

4. $6 \cdot 1\frac{2}{3}$ _____

5. $2\frac{3}{4} \cdot \frac{1}{4}$ _____

6. $14 \cdot \frac{1}{7}$ _____

7. $4\frac{1}{6} \cdot 12$ _____

8. $3\frac{1}{2} \cdot \frac{8}{9}$ _____

9. $7\frac{1}{5} \cdot \frac{10}{11}$ _____

10. $2\frac{1}{4} \cdot 1\frac{1}{2}$ _____

11. $2\frac{3}{8} \cdot 2\frac{1}{4}$ _____

12. $6\frac{3}{5} \cdot 2\frac{1}{3}$ _____

13. $3\frac{1}{3} \cdot 2\frac{2}{3}$ _____

14. $1\frac{5}{8} \cdot 4\frac{2}{5}$ _____

15. $9\frac{1}{9} \cdot 8\frac{1}{8}$ _____

16. $3\frac{1}{6} \cdot 1\frac{4}{5}$ _____

17. $10\frac{7}{8} \cdot 1\frac{5}{9}$ _____

18. $3\frac{6}{11} \cdot 12\frac{1}{2}$ _____

Name the reciprocal of each number. Show that the product of each number and its reciprocal is 1.

19. $\frac{4}{5}$ _____

20. $\frac{8}{15}$ _____

21. $\frac{1}{3}$ _____

22. 7 _____

23. $\frac{99}{100}$ _____

24. $2\frac{1}{7}$ _____

25. $1\frac{3}{8}$ _____

26. $5\frac{9}{10}$ _____

27. $3\frac{11}{12}$ _____

Estimate each quotient; then multiply by the reciprocal to find each quotient. Use a calculator to check the quotient.

28. $\frac{2}{3} \div \frac{1}{2}$ _____

29. $\frac{4}{5} \div \frac{1}{10}$ _____

30. $\frac{1}{6} \div \frac{1}{3}$ _____

31. $6 \div \frac{2}{3}$ _____

32. $\frac{1}{9} \div \frac{2}{3}$ _____

33. $\frac{7}{9} \div \frac{3}{4}$ _____

34. $\frac{5}{8} \div 8$ _____

35. $\frac{3}{4} \div \frac{1}{16}$ _____

36. $\frac{5}{6} \div \frac{5}{12}$ _____

37. $4 \div \frac{3}{4}$ _____

38. $6 \div 1\frac{2}{3}$ _____

39. $8 \div 5\frac{4}{5}$ _____

40. $2\frac{2}{5} \div 6$ _____

41. $3\frac{1}{2} \div 7$ _____

42. $\frac{7}{10} \div 2\frac{1}{5}$ _____

43. $4\frac{3}{8} \div \frac{3}{4}$ _____

44. $7\frac{2}{3} \div \frac{1}{3}$ _____

45. $8\frac{2}{5} \div \frac{7}{10}$ _____

46. $1\frac{1}{2} \div 2\frac{2}{3}$ _____

47. $1\frac{2}{3} \div 4\frac{1}{5}$ _____

48. $9\frac{1}{3} \div 2\frac{3}{4}$ _____

Practice
3.6 Exploring Ratios

Complete each proportion by using the property of equivalent ratios.

1. $\dfrac{2}{3} = \dfrac{4}{?}$ _____

2. $\dfrac{4}{5} = \dfrac{?}{15}$ _____

3. $\dfrac{9}{10} = \dfrac{?}{40}$ _____

4. $\dfrac{?}{169} = \dfrac{2}{13}$ _____

5. $\dfrac{14}{?} = \dfrac{42}{51}$ _____

6. $\dfrac{?}{24} = \dfrac{5}{6}$ _____

7. $\dfrac{3}{7} = \dfrac{27}{?}$ _____

8. $\dfrac{13}{14} = \dfrac{?}{56}$ _____

9. $\dfrac{16}{24} = \dfrac{8}{?}$ _____

10. $\dfrac{20}{48} = \dfrac{?}{12}$ _____

11. $\dfrac{32}{?} = \dfrac{4}{7}$ _____

12. $\dfrac{?}{35} = \dfrac{2}{5}$ _____

13. $\dfrac{3000}{730} = \dfrac{?}{73}$ _____

14. $\dfrac{1450}{?} = \dfrac{290}{300}$ _____

15. $\dfrac{2}{5} = \dfrac{1000}{?}$ _____

Out of 500 students surveyed, 320 have at least one brother, 325 have at least one sister, and 122 have no brothers or sisters. Write a ratio to describe each situation. Then write an equivalent ratio in simplest form.

16. the ratio of the number of students with at least one brother to the number of students surveyed _____

17. the ratio of the number of students with at least one sister to the number of students surveyed _____

18. the ratio of the number of students with no brothers or sisters to the number of students surveyed _____

Kelli puts 70 plants in her garden. She plants 16 tomato plants, 20 pepper plants, 4 zucchini plants, 10 cucumber plants, and 20 marigold plants. Write a ratio to describe each situation. Then write an equivalent ratio in simplest form.

19. the ratio of the number of tomato plants to the total number of plants _____

20. the ratio of the number of pepper plants to the total number of plants _____

21. the ratio of the number of zucchini plants to the total number of plants _____

22. the ratio of the number of cucumber plants to the total number of plants _____

23. the ratio of the number of marigold plants to the total number of plants _____

Practice
3.7 Percent

Use a percent bar to write each percent as a fraction in simplest form.

1. 30%

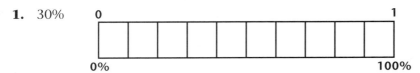

_____

2. 25%

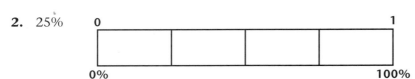

_____

3. 150%

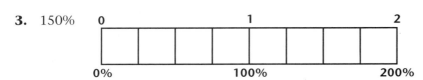

_____

What percent does each bar model represent?

4.

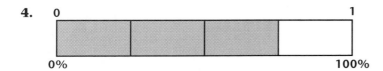

_____

5.

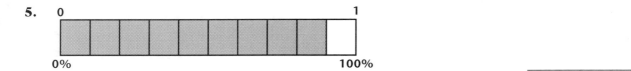

_____

6.

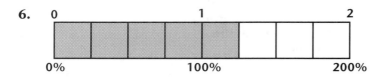

_____

Change each decimal to a percent.

7. 0.4 _____   **8.** 0.6 _____   **9.** 0.93 _____   **10.** 0.17 _____

11. 0.925 _____   **12.** 0.138 _____   **13.** 3.99 _____   **14.** 6.045 _____

Change each fraction to a percent.

15. $\frac{1}{4}$ _____   **16.** $\frac{5}{8}$ _____   **17.** $\frac{3}{10}$ _____   **18.** $\frac{3}{5}$ _____

19. $\frac{1}{3}$ _____   **20.** $\frac{1}{6}$ _____   **21.** $1\frac{9}{10}$ _____   **22.** $5\frac{11}{12}$ _____

 **Practice**

3.8 Experimental Probability

Two coins were flipped 20 times with the following results:

Trial	1	2	3	4	5	6	7	8	9	10	11	12	13	14	15	16	17	18	19	20
Coin 1	H	T	H	H	T	H	T	T	H	T	H	T	T	H	T	H	H	T	H	T
Coin 2	T	T	H	H	H	T	T	H	T	T	T	T	T	H	H	T	H	T	T	H

According to the data, find the following experimental probabilities:

1. Both coins are alike. _____

2. Both coins are heads. _____

3. At least one coin is heads. _____

4. Neither coin is heads. _____

5. Both coins are tails. _____

6. At least one coin is tails. _____

7. Neither coin is tails. _____

8. Both coins are different. _____

The command RAND generates a decimal value from 0 to 1, including 0 but not including 1. Describe the output of each command.

9. RAND*3 _____

10. INT(RAND*3) _____

11. INT(RAND*3)+1 _____

12. 10* (INT(RAND*3)+1) _____

13. 100*(INT(RAND*3)+1) _____

Write commands to generate random numbers from each of the following lists, where RAND generates a number from 0 to 1, including 0, but not including 1. If you can, check your results by using technology. Adapt the command to suit the computer or calculator that you are using.

14. 0, 1, 2, 3 _____

15. 0, 1, 2, 3, 4, 5, 6, 7 _____

16. 1, 2, 3 ,4 , 5, 6, 7 _____

17. 1, 2, 3, 4, 5, 6, 7, 8, 9, 10, 11, 12 _____

18. 3, 4, 5 _____

19. 7, 8, 9, 10, 11, 12 _____

20. 10, 11, 12, 13, 14, 15, 16 _____

Practice
3.9 Theoretical Probability

A bag contains 16 marbles: 10 blue (B), 4 red (R), and 2 green (G). One marble is randomly drawn from the bag. Use this information for Exercises 1–10.

1. $P(B)$ _____

2. $P(R)$ _____

3. $P(G)$ _____

4. $P(not\ B)$ _____

5. $P(not\ R)$ _____

6. $P(not\ G)$ _____

7. $P(B\ or\ R)$ _____

8. $P(R\ or\ G)$ _____

9. $P(G\ or\ B)$ _____

10. What is $P(B) + P(R) + P(G)$? _____

Find the theoretical probability for one roll of an ordinary 6-sided number cube.

11. $P(4)$ _____

12. $P(4\ or\ 5)$ _____

13. $P(even)$ _____

14. $P(odd)$ _____

15. $P(> 4)$ _____

16. $P(< 4)$ _____

Find the theoretical probability when two ordinary 6-sided number cubes are rolled.

17. $P(4\ and\ 5)$ _____

18. $P(4\ and\ 6)$ _____

19. $P(4\ and\ 4)$ _____

20. $P(both\ even)$ _____

21. $P(both\ odd)$ _____

22. $P(doubles)$ _____

A bag contains 16 marbles: 10 blue (B), 4 red (R), and 2 green (G). Two marbles are randomly drawn from the bag. Use the grid to find the following probabilities:

		First marble returned (independent event)	First marble not returned (dependent event)
23.	$P(B,\ then\ B)$		
24.	$P(B,\ then\ R)$		
25.	$P(B,\ then\ G)$		
26.	$P(R,\ then\ B)$		
27.	$P(R,\ then\ R)$		
28.	$P(R,\ then\ G)$		
29.	$P(G,\ then\ B)$		
30.	$P(G,\ then\ R)$		
31.	$P(G,\ then\ G)$		
32.	SUM		

Practice
4.1 Lines and Angles

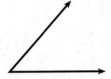

Use a protractor to measure each angle.

1.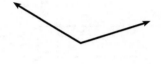

_____

2.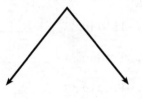

_____

3.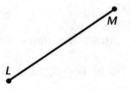

_____

Name each figure in all possible ways.

4.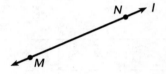

_____

5.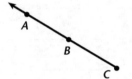

_____

6.

_____

Use a protractor to draw angles with the following measures.
Label each as either acute, obtuse, right, or straight.

7. 55°

8. 100°

9. 180°

10. 125°

11. 90°

12. 25°

Practice

4.2 Exploring Angles

In the figure shown, m∠1 = 40.5°. Use the figure for Exercises 1–10.

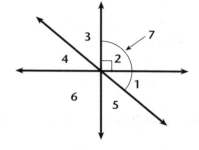

1. Find m∠2. _____

2. Find m∠4. _____

3. Find m∠3. _____

4. Find m∠5. _____

5. Find m∠6. _____

6. Find m∠7. _____

7. Name three pairs of vertical angles. _____

8. Name two right angles. _____

9. Name two pairs of complementary angles. _____

10. Name two pairs of supplementary angles. _____

Use the figure shown for Exercises 11–22.

11. Find m∠AFB. _____

12. Find m∠AFE. _____

13. Find m∠CFB. _____

14. Find m∠CFA. _____

15. Find m∠CFE. _____

16. Name three acute angles. _____

17. Name three obtuse angles. _____

18. Name one pair of vertical angles. _____

19. Name two right angles. _____

20. Name one pair of complementary angles. _____

21. Name two straight angles. _____

22. Name two pairs of supplementary angles. _____

Practice
4.3 Exploring Parallel Lines and Triangles

In the figure shown, *l* || *m* and m∠5 = 110°.
Find the measure of each angle.

1. m∠3 _____   2. m∠2 _____

3. m∠1 _____   4. m∠6 _____

5. m∠7 _____   6. m∠8 _____

7. m∠4 _____

8. List all pairs of alternate interior angles. _____

9. List all pairs of alternate exterior angles. _____

10. List all pairs of consecutive interior angles. _____

11. Explain why ∠3 and ∠4 are congruent. _____

12. Name 10 pairs of supplementary angles. _____

_____

Classify each of the following triangles according to its angles
and according to its sides.

13. 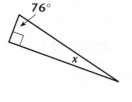   14. 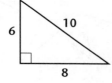   15.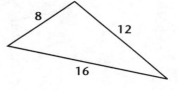

_____   _____   _____

Find the missing angle measures in each triangle.

16.    17.   18.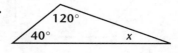

_____   _____   _____

Practice
4.4 Exploring Polygons

Label all of the side lengths and all of the angle measures for each parallelogram.

1.
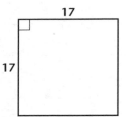

2.
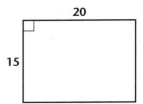

3.
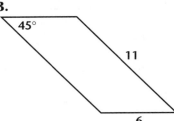

4.
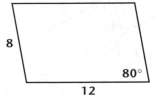

5.
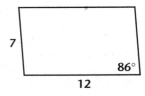

6.
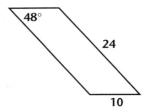

Classify each polygon as concave or convex, name the polygon, and find the sum of its angle measures.

7.
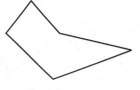

8.

9.

_____

_____

_____

10.
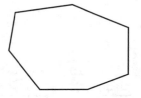

11.

12.
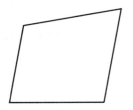

_____

_____

_____

Practice

4.5 Exploring Perimeter and Area

Estimate the perimeter and area of each figure.

1.

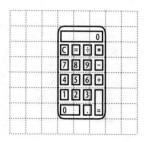

_____

2.

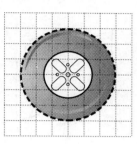

_____

3.

_____

Use formulas to compute the perimeter and area of each rectangle.

4. $l = 13$ meters, $w = 12$ meters _____

5. $l = 2.5$ feet, $w = 2.25$ feet _____

6. $l = 9\frac{1}{8}$ inches, $w = 3\frac{5}{8}$ inches _____

7. $l = 90$ feet, $w = 90$ feet _____

8. $l = 1.8$ miles, $w = 0.2$ miles _____

9. $l = 3\frac{1}{3}$ yards, $w = 1\frac{1}{3}$ yards _____

10. $l = 2\frac{1}{2}$ inches, $w = \frac{9}{16}$ inches _____

11. $l = 12.9$ centimeters, $w = 8.5$ centimeters _____

Use formulas for perimeter and area to compute the length and area of each rectangle.

12. $P = 24$ meters, $w = 3.75$ meters _____

13. $P = 43$ centimeters, $w = 3.4$ centimeters _____

14. $P = 21\frac{1}{4}$ inches, $w = 2\frac{3}{8}$ inches _____

15. $P = 33$ meters, $w = 4.25$ meters _____

Practice
4.6 Exploring Area Formulas

Use the grid to find the area of each figure.

1.

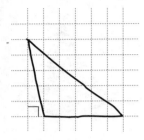

_____

2.

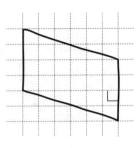

_____

3.

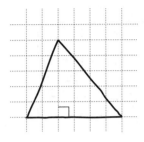

_____

Find the area of each figure.

4.

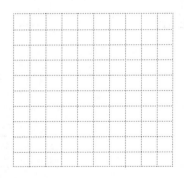

_____

5.

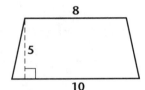

_____

6.

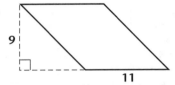

_____

On each grid provided, sketch a picture of each parallelogram with the indicated base and height. Then use the formula to compute the area.

7. $b = 6$ centimeters, $h = 4.5$ centimeters

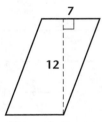

_____

8. $b = 5\frac{1}{2}$ feet, $h = 6\frac{1}{4}$ feet

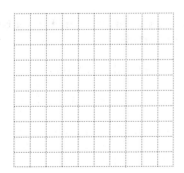

_____

Practice
4.7 Square Roots and the "Pythagorean" Right-Triangle Theorem

Estimate each square root to the nearest whole number and to the nearest tenth. Then use a calculator to determine each square root to the nearest hundredth.

1. $\sqrt{7}$ _____   **2.** $\sqrt{14}$ _____   **3.** $\sqrt{24}$ _____

4. $\sqrt{28}$ _____   **5.** $\sqrt{70}$ _____   **6.** $\sqrt{55}$ _____

7. $\sqrt{90}$ _____   **8.** $\sqrt{105}$ _____   **9.** $\sqrt{83}$ _____

10. $\sqrt{140}$ _____   **11.** $\sqrt{119}$ _____   **12** $\sqrt{224}$ _____

13. $\sqrt{66}$ _____   **14.** $\sqrt{220}$ _____   **15.** $\sqrt{34}$ _____

16. $\sqrt{150}$ _____   **17.** $\sqrt{67}$ _____   **18.** $\sqrt{149}$ _____

The lengths of the sides of a triangle are given. Use the converse of the "Pythagorean" Right-Triangle Theorem to determine if the triangle is a right triangle.

19. 9, 12, 15 _____   **20.** 2, 5, 7 _____   **21.** 10, 15, 18 _____

22. 2.5, 6, 6.5 _____   **23.** 3, 4, 5 _____   **24.** 7, 9, 10 _____

25. 10, 24, 26 _____   **26.** 8, 15, 17 _____   **27.** 15, 20, 25 _____

28. 8, 12, 13 _____   **29.** 5, 15, 18 _____   **30.** 12, 35, 37 _____

31. 10, 10.5, 14.5 _____   **32.** 16, 30, 34 _____   **33.** 25, 30, 36 _____

34. 17, 19, 35 _____   **35.** 14, 7, 12 _____   **36.** 1.5, 2, 2.5 _____

For Exercises 37–39, use the "Pythagorean" Right-Triangle Theorem to find the unknown side of each right triangle.

37.

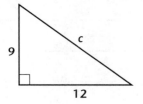

38.

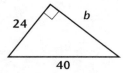

39.

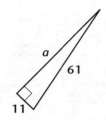

_____   _____   _____

Practice
4.8 Scale Factors and Similarity

Use the scale factor to find the length and width of each rectangle. Then sketch a picture of each rectangle.

	Scale factor	Length	Width
	original	6 cm	14 cm
1.	2		
2.	3		
3.	4		
4.	$\frac{1}{4}$		
5.	0.5		
6.	1.4		
7.	80%		
8.	$\frac{3}{2}$		

The right triangles described in the table are similar right triangles. Find the scale factor and the length of the unknown leg.

	Leg	Leg	Scale factor
	6	10	original
9.	18		
10.		40	
11.	1		
12.		5	
13.	2		

Practice
5.1 Adding Expressions

Add.

1. $(2y + 3) + (y - 2)$ _____

2. $(3x + 5) + (6x - 1)$ _____

3. $(5a + 6) + (2a - 4)$ _____

4. $(5z - 3) + (7 - z)$ _____

5. $(3r + 5) + (r + 1)$ _____

6. $(3 - 4y) + (3y + 6)$ _____

7. $(5 + 2w) + (-6w + 1)$ _____

8. $(6x - 2y) + (x - y)$ _____

9. $(2n + 3) + (-3 - 2n)$ _____

10. $(-2m - 3) + (3m + 5.3)$ _____

11. $(0.5t - 1) + (1.4t + 2)$ _____

12. $(0.2d + 0.5g) + (1.8d - 0.5g)$ _____

13. $(8.2b - 7.6) + (3.2 - 4.5b)$ _____

14. $(-3x - 9.8) + (3 - 6.2x)$ _____

15. $\left(\frac{x}{2} + 2\right) + \left(\frac{x}{4} + 1\right)$ _____

16. $\left(\frac{x}{8} + \frac{3}{4}\right) + \left(\frac{x}{2} - \frac{1}{8}\right)$ _____

17. $\left(\frac{4v}{5} + \frac{1}{2}\right) + \left(\frac{3v}{5} + \frac{2}{3}\right)$ _____

18. $\left(\frac{3n}{10} + 2n\right) + \left(\frac{4n}{5} - n\right)$ _____

19. $(3a + 2b + c) + (a - b - c)$ _____

20. $(5d - 6f + 3g) + (4f - g + d)$ _____

21. $(4x + y) + (y + z) + (x - 6z)$ _____

22. $(m - q) + (3n + 4q) + (q - 6m)$ _____

23. $(6r - 5s) + (3 - 2r) + (6 - 3s)$ _____

24. $(m + n - 4p) + (3m + n) + (n - p)$ _____

Use the Distributive Property to write an equivalent expression for each.

25. $2(3 + 8)$ _____

26. $7(x + 2)$ _____

27. $4(y + z)$ _____

28. $b(c + d)$ _____

29. $(j - k)h$ _____

30. $(1 - t)r$ _____

31. $(2 + t)5$ _____

32. $n(m - 8)$ _____

33. $r(t - s)$ _____

34. $(5 \cdot 2 + 3 \cdot 2)$ _____

35. $3w + 3 \cdot 9$ _____

36. $hj + hk$ _____

37. $4r + 4s$ _____

38. $mn - 3n$ _____

Practice
5.2 Subtracting Expressions

Find the opposite of each expression.

1. 13 _____   **2.** -25 _____   **3.** -32 _____   **4.** $2x$ _____

5. $-5w$ _____   **6.** a _____   **7.** $-v$ _____   **8.** $a + 5$ _____

9. $r - 2$ _____   **10.** $3x + 1$ _____   **11.** $-2 - q$ _____   **12.** $9a + b$ _____

13. $-10t + 1$ _____   **14.** $-x + 2z$ _____   **15.** $-13 - t$ _____   **16.** $4a - 4b$ _____

17. $a - b + c$ _____   **18.** $-4x + 3y - z$ _____   **19.** $3w - 4z + 2$ _____

20. $x + y - 5z$ _____   **21.** $-7a - 6b - 4c$ _____   **22.** $m - 8n - 4p$ _____

Perform the indicated operations.

23. $2a - a$ _____   **24.** $6f - 4f$ _____

25. $6r - 3r$ _____   **26.** $2d - 7d$ _____

27. $5a - (3a + 1)$ _____   **28.** $3 - (3d - 2d)$ _____

29. $6x - (2x + 5)$ _____   **30.** $9z - (3 - 6z)$ _____

31. $(x + y) - (3x + y)$ _____   **32.** $5x - (3x + 4)$ _____

33. $(3a + 2b) - (3a - b)$ _____   **34.** $(6p - 3q) - (-6p)$ _____

35. $(6 - 2c) - (2c + 2)$ _____   **36.** $10 - (r - 10)$ _____

37. $(4q + 2) - (2q - 3) + (3q - 1)$ _____   **38.** $(2m - 2) - (3m - 2) + (m - 1)$ _____

39. $(m + n - p) - (2m + 3n + 4p)$ _____   **40.** $(7w + 4x) - (3w - z) + (x + z)$ _____

41. $(4 - w) - (w - t) - (w - t)$ _____   **42.** $(5a - 3c) - (b + c) + (3b - 2c)$ _____

43. $(3 - r) + (4r - 3s + 2) - (1 - s)$ _____

44. $(3x - 5y) - (-2x - 3y - z) + (y - z)$ _____

Tell whether each statement is true or false or cannot be determined. Then explain why.

45. $-a - (b + c) = -a - b + c$ _____

46. $-a - (b - c) = -a - b + c$ _____

47. $-(a + b)$ is a negative number. _____

Practice
5.3 Exploring Addition and Subtraction Equations

Write the equation modeled by each set of tiles.

1.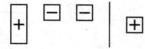

_____

2.

_____

3.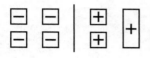

_____

4.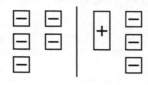

_____

Use algebra tiles to set up each equation.

5. $x - 4 = 2$ _____

6. $-3 = x + 2$ _____

Solve each equation.

7. $x - 1 = 4$ _____   **8.** $t + 5 = 1$ _____   **9.** $h - 4 = 6$ _____

10. $x + 7 = 1$ _____   **11.** $y - 5 = -2$ _____   **12.** $g - 3 = 15$ _____

13. $-5 = d - 2$ _____   **14.** $7 = 4 + t$ _____   **15.** $n - 4 = -8$ _____

16. $(-8) + m = 8$ _____   **17.** $-8 = y - 5$ _____   **18.** $6 = y - 4$ _____

19. $(-5) + y = 9$ _____   **20.** $7 = 4 - t$ _____   **21.** $-2 = r + 4$ _____

22. $15 = v + (-3)$ _____   **23.** $b - 2 = -2$ _____   **24.** $f + 5 = 5$ _____

25. $14 = -5 + y$ _____   **26.** $8 = x - 5$ _____   **27.** $-9 + r = 10$ _____

28. $-8 = -4 + w$ _____   **29.** $x + 19 = -10$ _____   **30.** $s - 9 = -4$ _____

31. $-7 + t = -6$ _____   **32.** $4 = 6 - x$ _____   **33.** $13 = -9 + r$ _____

Practice

5.4 Addition and Subtraction Equations

Solve each equation. You may use algebra tiles.

1. $x + 1 = 5$ _____

2. $x - 7 = 1$ _____

3. $x + 3 = -2$ _____

4. $x - 4 = -1$ _____

5. $x + 1 = -8$ _____

6. $x - 3 = -1$ _____

7. $x + 5 = -2$ _____

8. $x - 3 = 4$ _____

9. $x + 4 = -4$ _____

State which property you would use to solve each equation. Then solve.

10. $x - 10 = 15$ _____

11. $x + 14 = -25$ _____

12. $45 + r = 12$ _____

13. $y + 80 = -18$ _____

14. $-44 + a = 10$ _____

15. $z + 250 = -100$ _____

16. $y - 12 = 78$ _____

17. $r - 275 = 180$ _____

18. $x + 6.26 = 7.26$ _____

19. $x - 3.6 = 7$ _____

20. $8.9 = a - 6$ _____

21. $r + 6.5 = 10.9$ _____

22. $y - \frac{3}{4} = \frac{1}{8}$ _____

23. $x + \frac{1}{5} = \frac{3}{10}$ _____

Assign a variable and write an equation for the situation below. Then solve the equation.

24. John bought a 60¢ donut and 35¢ cup of coffee. How much did John leave for a tip and taxes if he spent a total of $1.25?

_____

Practice
5.5 Exploring Polynomials

Simplify each polynomial by combining like terms. Check by substituting two different values for x. If the polynomial is already simplified, write *simplified*.

1. $4x - 3x + 2$ _____

2. $3 + 5x - 3$ _____

3. $7x + 5 - 6x$ _____

4. $-7x + 2 - 7x$ _____

5. $x^2 + 2x + 3x^2$ _____

6. $6x - 7x^2 - 7x + 2$ _____

7. $5x + 3x - 2 + 4x^2$ _____

8. $4x - 1 - 5x - 2$ _____

9. $9x + 2 - 9x^2$ _____

10. $7x - (4x^2 - 3)$ _____

11. $-3x - 3x^2 - (4x + 5)$ _____

12. $4 - 8x^2 - (x^2 + 6x)$ _____

Simplify each binomial addition expression. Check by substituting two different values for x.

13. $(3x + 1) + (2x - 3)$ _____

14. $(7x - 1) + (-5x + 3)$ _____

15. $(8x - 2) + (-6x + 3)$ _____

16. $(3x - 3) + (9 - 5x)$ _____

17. $(7 - 3x) + (2x - 3)$ _____

18. $(-2 + 9x) + (4 + 3x)$ _____

19. $(3x - 2) + (-4x^2 + x)$ _____

20. $(5x + 5) + (-3x^2 + 2)$ _____

21. $(-4x - 3) + (-5 - x)$ _____

22. $(-x + 2) + (-3 + x)$ _____

23. $(-6x^2 - 3x) + (4x - 1)$ _____

24. $(-8x - 1) + (-4x - 8x^2)$ _____

Simplify each binomial subtraction expression. Check by substituting two different values for x.

25. $(3x - 2) - (4x - 4)$ _____

26. $(2x - 4) - (7x - 1)$ _____

27. $(3 - 5x) - (6x - 3)$ _____

28. $(-x - 2) - (-2x - 2)$ _____

29. $(-5x - 4) - (-6 - x)$ _____

30. $(7x + 4) - (-4x - 7)$ _____

31. $(x^2 + 1) - (2x^2 - 5)$ _____

32. $(3x^2 - 2) - (5x^2 + 6)$ _____

33. $(7x^2 - 7) - (6x^2 - 3)$ _____

34. $(-x^2 - 10) - (5 - 3x^2)$ _____

35. $(-x^2 - x) - (-4x^2 + 3)$ _____

36. $(-x - 4) - (x^2 + 3x)$ _____

Practice
5.6 Exploring Inequalities

Graph each inequality on the number line provided.

1. $x < 5$

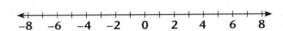

2. $x \geq -2$

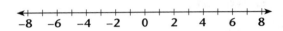

3. $x > -1$

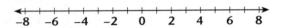

4. $x \leq -6$

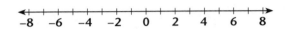

5. $3 \geq x$

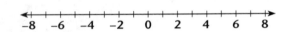

6. $-3 < x$

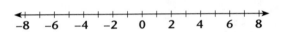

Determine whether the number following each inequality is a solution of the inequality.

7. $x + 2 > 1; 3$ _____

8. $g - 3 \leq 2; 1$ _____

9. $m - 3 \geq -2; 1$ _____

10. $x - 4 \leq 7; 10$ _____

11. $f - 9 > 15; -9$ _____

12. $-7 + z < -1; -1$ _____

13. $2 + y < 14; 10$ _____

14. $7 \leq t + 2; -5$ _____

15. $2 \leq -5 + t; -4$ _____

Use any method to solve each inequality.

16. $x + 7 \geq 3$ _____

17. $y + 2 \leq -3$ _____

18. $15 + z \geq 10$ _____

19. $x + 0.8 \leq 2.8$ _____

20. $5 + r < -2$ _____

21. $h - 7 \leq -3$ _____

22. $5 \leq 12 + x$ _____

23. $6 > m - 13$ _____

24. $r + 1.2 \leq -4.5$ _____

25. $12 > t - 6$ _____

26. $8 < b - 5$ _____

27. $12 + y \leq -4$ _____

28. $y + 450 < 550$ _____

29. $7000 \geq r - 5000$ _____

30. $500 + w \leq 1000$ _____

 **Practice**

5.7 Solving Related Inequalities

State whether each inequality is true or false.

1. $5 > 8 - 2$ _____

2. $-1 - 1 > 0$ _____

3. $-1 \leq 5 - 6$ _____

4. $-1 + 2 \leq 3$ _____

5. $-4 - 2 < 4$ _____

6. $8 > 6 - 4$ _____

7. $7 - 9 \leq 1$ _____

8. $12 < 9 - 10$ _____

9. $3 \leq -8 + 11$ _____

Solve each inequality.

10. $x + 5 \geq -3$ _____

11. $t - 5 < -2$ _____

12. $8 + y \leq -1$ _____

13. $c + \frac{1}{2} < 1$ _____

14. $q - \frac{1}{3} \geq 3$ _____

15. $x + 0.8 \geq 1$ _____

16. $0.75 > -0.5 + d$ _____

17. $x + 4.9 \leq 0.45$ _____

18. $3.35 \geq -4.85 + n$ _____

19. $y + \frac{3}{4} > \frac{1}{8}$ _____

20. $\frac{5}{6} \leq x + \frac{2}{3}$ _____

21. $\frac{2}{5} > t - \frac{7}{10}$ _____

Write an inequality that describes the points on each number line.

22.
```
<-+--+--+--⊕--+--+--+--+--+->
 -8  -6  -4  -2  0   2   4   6   8
```
_____

23.
```
<-+--+--+--+--+--+--●--+--+->
 -8  -6  -4  -2  0   2   4   6   8
```
_____

24.
```
<-+--+--⊕--+--⊕--+--+--+--+->
 -8  -6  -4  -2  0   2   4   6   8
```
_____

25.
```
<-+--+--●--+--+--●--+--+--+->
 -8  -6  -4  -2  0   2   4   6   8
```
_____

26.
```
<-+--+--+--+--⊕--+--+--●--+->
 -8  -6  -4  -2  0   2   4   6   8
```
_____

27.
```
<-+--+--●--+--+--⊕--+--+--+->
 -8  -6  -4  -2  0   2   4   6   8
```
_____

28.
```
<-+--●--+--+--+--+--+--+--+->
 -8  -6  -4  -2  0   2   4   6   8
```
_____

Practice
6.1 Multiplying and Dividing Expressions

Evaluate 16b for the following values of b:

1. -7 _____   **2.** 3.2 _____   **3.** 5 _____   **4.** $\frac{1}{4}$ _____

Evaluate 4a + 1 for the following values of a:

5. 30 _____   **6.** 2.7 _____   **7.** -4.5 _____   **8.** $\frac{2}{5}$ _____

Simplify each expression.

9. $2 \cdot 7x$ _____   **10.** $-4x \cdot 3$ _____   **11.** $4x \cdot 3x$ _____

12. $-4(2x + 3)$ _____   **13.** $-2.3y \cdot 2y$ _____   **14.** $-12w \div 2$ _____

15. $3y \cdot -2 + 4y \cdot 2$ _____   **16.** $-3y \cdot -4.5y$ _____   **17.** $-5(7 - 4y)$ _____

18. $-36z \div -4$ _____   **19.** $-15x \cdot -3x$ _____   **20.** $-4(6y - 4)$ _____

21. $-3(2 - 7x)$ _____   **22.** $3x - (2x - 5)$ _____   **23.** $w - 3(2 - w)$ _____

24. $x - 5(x + 2)$ _____   **25.** $3(2 - y) + 5y$ _____   **26.** $-(2x - 6) - 8$ _____

27. $-(-x - y) - 4y$ _____   **28.** $\frac{6 + 12w}{2}$ _____   **29.** $\frac{10 - 5x}{-5}$ _____

30. $\frac{-4 + 6x}{2}$ _____   **31.** $\frac{-18w + 12}{-6}$ _____   **32.** $\frac{-32 - 24y}{-4}$ _____

33. $(7z - 1) - 5(z + 2)$ _____   **34.** $2(x - 2) - 3(3 - x)$ _____

35. $4(-x - w) - (2x - 3w)$ _____   **36.** $2(3x - 1) - 4(x + 2)$ _____

37. $-(x - y) + 2(5x - 3y)$ _____   **38.** $(3x + 2y - 3) - (4x - 5y)$ _____

A computer consultant charges \$50 per hour. How much would the consultant charge for

39. 3 hours? _____   **40.** 7.5 hours? _____   **41.** t hours? _____

The telephone company charges \$40 per hour for repair work, plus a \$25 service charge per job. How much would a customer be charged for a job that takes

42. 2 hours? _____   **43.** 3.5 hours? _____   **44.** h hours? _____

Practice
6.2 Multiplication and Division Equations

State the property needed to solve each equation. Then solve it.

1. $\frac{x}{36} = 6$ _____

2. $5.75p = 46$ _____

3. $8 = -64y$ _____

4. $x + \frac{1}{3} = 3$ _____

5. $6b = 54$ _____

6. $\frac{r}{-5} = 5$ _____

7. $y - \frac{2}{5} = -10$ _____

8. $-16n = -128$ _____

Solve each equation.

9. $658b = 2632$ _____

10. $-4g = 36$ _____

11. $x + 65 = 35$ _____

12. $x + \frac{2}{3} = 24$ _____

13. $y - 658 = 2532$ _____

14. $\frac{c}{44} = -44$ _____

15. $\frac{x}{0.07} = 5$ _____

16. $8t = -35$ _____

17. $0.66p = 4.62$ _____

18. $g + 4200 = 580$ _____

19. $\frac{b}{20} = -20$ _____

20. $b + 87 = 59$ _____

21. $\frac{m}{-15} = 0$ _____

22. $x - \frac{1}{6} = 3$ _____

23. $4.4 = 2.2t$ _____

24. $n - 3500 = 4800$ _____

25. $-4v = 68.8$ _____

26. $\frac{x}{-2} = -125$ _____

27. $-480 = -24z$ _____

28. $d - \frac{3}{5} = 10$ _____

29. $0.25q = 25$ _____

30. $\frac{w}{0.5} = -15$ _____

31. $-1 = \frac{p}{-555}$ _____

32. $-495 = 99y$ _____

Solve each formula for the variable indicated.

33. $A = \frac{1}{2}bh$ for b _____

34. $P = 2l + 2w$ for w _____

35. $v = \frac{d}{t}$ for d _____

 **Practice**

6.3 Exploring Products and Factors

Write the factors and the product modeled by the tiles.

1.

×	+	−	−	−
+	+	−	−	−

2.

×	+	+	+	+
+	+	+	+	+
+	+	+	+	+

_____   _____

Draw a tile model to show the area of a rectangle represented by each product.

3. $x(x + 4)$ **4.** $2x(2x + 1)$

Use the Distributive Property to find each product.

5. $6(x + 2)$ _____   **6.** $6x(x + 2)$ _____   **7.** $b^2(b + 3)$ _____

8. $3y(y^2 - 1)$ _____   **9.** $3y^2(y^2 + 1)$ _____   **10.** $3(m^2 + 8)$ _____

11. $12x(2x + 3)$ _____   **12.** $5y(2y^2 - 6)$ _____   **13.** $z^2(z^2 - 5z)$ _____

Write each polynomial in factored form.

14. $6p - 24$ _____   **15.** $9y + 12$ _____   **16.** $r^2 + 6r$ _____

17. $2p^2 - 12p$ _____   **18.** $6w^3 - 4w^2$ _____   **19.** $-6t^2 - 5t$ _____

20. $8y^3 + 40y^2$ _____   **21.** $9a - 9b$ _____   **22.** $-14t + 21s$ _____

23. $32x^4 + 16x^2$ _____   **24.** $24m^4 - 32m^3$ _____   **25.** $-15n^2 - 10n$ _____

 **Practice**

6.4 Rational Numbers

Write the reciprocal of each number.

1. -5 _____

2. $\frac{5}{6}$ _____

3. $-\frac{2}{3}$ _____

4. 6 _____

5. $\frac{-1}{14}$ _____

6. $\frac{12}{5}$ _____

7. 2 _____

8. $7\frac{1}{8}$ _____

Solve each equation.

9. $\frac{4}{5}y = 20$ _____

10. $\frac{1}{-8}x = -4$ _____

11. $\frac{w}{-5} = 6$ _____

12. $\frac{p}{-12} = -3$ _____

13. $\frac{2}{3}z = -12$ _____

14. $-\frac{r}{7} = 9$ _____

15. $\frac{-3}{5}t = 8$ _____

16. $\frac{-p}{-2} = 14$ _____

17. $\frac{z}{-16} = -20$ _____

18. $\frac{-x}{9} = -6$ _____

19. $\frac{5}{6}x = 5$ _____

20. $\frac{r}{3} = -1.8$ _____

Solve each proportion.

21. $\frac{x}{6} = \frac{2}{3}$ _____

22. $\frac{-x}{5} = \frac{4}{5}$ _____

23. $\frac{x}{7.5} = -\frac{1}{5}$ _____

24. $\frac{x}{30} = \frac{3}{2}$ _____

25. $\frac{-x}{5} = \frac{3}{4}$ _____

26. $\frac{y}{3.2} = \frac{-2}{4}$ _____

27. $\frac{x}{20} = \frac{-5}{4}$ _____

28. $\frac{x}{7} = \frac{4}{14}$ _____

29. $\frac{x}{3} = \frac{5}{6}$ _____

30. $\frac{-x}{5} = \frac{6}{3}$ _____

31. $\frac{y}{4} = \frac{3}{8}$ _____

32. $\frac{x}{-15} = \frac{2}{5}$ _____

33. $\frac{-x}{-10} = \frac{2}{15}$ _____

34. $\frac{t}{-1.4} = \frac{-5}{2}$ _____

35. $\frac{-b}{2.4} = \frac{3}{-4}$ _____

36. $\frac{y}{-5} = \frac{-14}{-1.5}$ _____

37. $\frac{-c}{10.8} = \frac{-2}{3}$ _____

38. $\frac{k}{-1.2} = \frac{-2.4}{5}$ _____

 **Practice**
6.5 Exploring Proportion Problems

Solve each proportion.

1. $\frac{12}{18} = \frac{x}{36}$ _____

2. $\frac{30}{27} = \frac{90}{x}$ _____

3. $\frac{n}{26} = \frac{15}{78}$ _____

4. $\frac{40.5}{t} = \frac{5}{6}$ _____

5. $\frac{m}{35} = \frac{12}{100}$ _____

6. $\frac{1.5}{6} = \frac{8}{r}$ _____

7. $\frac{45.2}{30} = \frac{f}{12}$ _____

8. $\frac{15}{90.5} = \frac{25}{x}$ _____

9. $\frac{n}{35.2} = \frac{14}{25.8}$ _____

10. $\frac{50.9}{16} = \frac{3}{j}$ _____

11. $\frac{r}{85} = \frac{1.5}{30}$ _____

12. $\frac{7}{16.6} = \frac{n}{14}$ _____

Determine if each statement is a true proportion.

13. $\frac{14}{3} = \frac{70}{15}$ _____

14. $\frac{7}{25} = \frac{3.5}{50}$ _____

15. $\frac{9}{32} = \frac{12}{40}$ _____

16. $\frac{10}{60} = \frac{25}{150}$ _____

17. $\frac{13}{24} = \frac{11}{35}$ _____

18. $\frac{7}{18} = \frac{21}{6}$ _____

19. $\frac{32}{42} = \frac{20}{21}$ _____

20. $\frac{19}{30} = \frac{9.5}{15}$ _____

21. $\frac{9}{12} = \frac{21}{28}$ _____

Rearrange the numbers to write three more true proportions.

22. $\frac{4}{5} = \frac{12}{15}$ _____

23. $\frac{7}{12} = \frac{14}{24}$ _____

24. $\frac{6}{21} = \frac{8}{28}$ _____

25. $\frac{32}{18} = \frac{48}{27}$ _____

26. $\frac{8}{3} = \frac{40}{15}$ _____

27. $\frac{42}{36} = \frac{35}{30}$ _____

28. If Dana spends $160 on 5 concert tickets, how much would 3 tickets cost? _____

29. Myron bought 4 oranges for $1.40. How much would 9 oranges cost? _____

30. A recipe uses 2 cups of flour and makes 24 muffins. How many cups of flour are needed to make 30 muffins? _____

Practice
6.6 Solving Problems Involving Percent

Write each percent as a decimal.

1. 15% _____   **2.** 3.2% _____   **3.** 0.7% _____   **4.** 6% _____

Write each decimal as a percent.

5. 0.62 _____   **6.** 0.041 _____   **7.** 0.002 _____   **8.** 5.00 _____

Draw a percent bar to model each problem.

9. Find 45% of 90.   **10.** What percent of 30 is 3?

Estimate each answer as more or less than 50 or 50%.

11. Find 28% of 60. _____   **12.** 15 is 35% of what number? _____

13. Find 125% of 65. _____   **14.** What percent of 50 is 125? _____

Find each answer.

15. What is 20% of 30? _____   **16.** What is 120% of 70? _____

17. 3 is what percent of 50? _____   **18.** 45 is what percent of 500? _____

19. 15 is 30% of what number? _____   **20.** 12 is 40% of what number? _____

21. What is 200% of 40? _____   **22.** What is 28% of 130? _____

23. A sweater is marked down from an original price of $45 to $33.75. By what percent has the original price of the sweater been marked down?   _____

24. A teacher says that 20% of your grade will be based on your portfolio. If there are a total of 400 points possible, how many points can you earn for your portfolio?   _____

25. The school newspaper reported that 32% of the student body pre-registered for classes. If the student body consists of 2000 students, how many of them pre-registered?   _____

Practice

7.1 Solving Two-Step Equations

Solve each equation.

1. $3x + 7 = 46$ _____

2. $8y - 5 = 43$ _____

3. $-4z + 7 = -15$ _____

4. $-4y + 28 = 72$ _____

5. $9x + 42 = 69$ _____

6. $-7 - 15w = 23$ _____

7. $-6x - 12 = 42$ _____

8. $9x - 36 = 0$ _____

9. $4x - 7 = 5$ _____

10. $5 - 3z = 32$ _____

11. $-4y + 2 = 29$ _____

12. $9 = 3 + 5y$ _____

13. $-16 = 3x - 1$ _____

14. $34 = 8 + 2z$ _____

15. $2(x - 4) = 8$ _____

16. $6(4 - 3x) = -48$ _____

17. $3 = -3(y + 5)$ _____

18. $2(3z - 10) = 40$ _____

19. $-24 = 3(-2y - 7)$ _____

20. $5z - 4 - 3z = 14$ _____

21. $8 - 3x + 5 + 5x = 7$ _____

22. $6z + 8 + 2z = 10$ _____

23. $4 = -7y + 13 + 9y + 1$ _____

24. $0 = 1 - 8z + 6 - 6z$ _____

25. $2x + 3 = -\frac{1}{2}$ _____

26. $\frac{3}{4}y - 3 = 9$ _____

27. $4y + \frac{2}{3} = 1$ _____

28. $2z - \frac{7}{10} = \frac{1}{2}$ _____

29. $4 + \frac{1}{4} + 6x = -\frac{11}{20}$ _____

30. $3\left(x - \frac{1}{4}\right) = \frac{3}{4}$ _____

31. $\frac{3}{4}x + 1 = 10$ _____

32. $1 + \frac{2}{3}y = 27$ _____

33. $\frac{3}{8} - \frac{1}{4}x = \frac{1}{16}$ _____

34. $\frac{1}{2}z - \frac{1}{2} = 5$ _____

35. $9z + 2 = \frac{1}{2}$ _____

36. $\frac{5}{2} = \frac{1}{4}y + 3$ _____

37. $\frac{2}{3} + 2 + \frac{3}{5} + x = \frac{4}{15}$ _____

38. $7x - 12 = -\frac{4}{5}$ _____

Practice
7.2 Solving Multistep Equations

Solve each equation.

1. $4x + 7 = 3x + 18$ _____

2. $5y - 5 = 7y - 3$ _____

3. $4a - 6 = -2a + 14$ _____

4. $4m - 5 = 3m + 7$ _____

5. $5x - 7 = 3x + 2$ _____

6. $10y + 10 = 4 - 4y$ _____

7. $13 - 8v = 5v + 2$ _____

8. $7 - 5y = 4y - 2$ _____

9. $2 + 3y = 4y - 1$ _____

10. $-7 - 3z = 8 + 2z$ _____

11. $7w - 19 = 5w - 5$ _____

12. $28 + 2a = 5a + 7$ _____

13. $5x + 32 = 8 - x$ _____

14. $m - 12 = 3m + 4$ _____

15. $2(x + 1) = 3x - 3$ _____

16. $5(3x + 5) = 4x - 8$ _____

17. $2r - 4 = 2(6 - 7r)$ _____

18. $8y - 3 = 5(2y + 1)$ _____

19. $2z - 5(z + 2) = -8 - 2z$ _____

20. $5t - 2(5 + 4t) = 3 + t - 8$ _____

21. $15n + 25 = 2(n - 7)$ _____

22. $4y + 2 = 3(6 - 4y)$ _____

23. $2(3x - 1) = 3(x + 2)$ _____

24. $9y - 8 + 4y = 7y + 16$ _____

25. $14d - 22 + 5d = 12d - 8$ _____

26. $23x + 34 = 23 - 12x + 7x$ _____

27. $29 - 3s = 23(2s - 3)$ _____

28. $12 - 5(2w - 13) = 3(2w - 5)$ _____

29. $8 + 5(3q - 4) = 7(q - 12)$ _____

30. $2(y + 2) + y = 19 - (2y + 3)$ _____

31. $0.3w - 4 = 0.8 - 0.2w$ _____

32. $2.1z = 1.2z - 9$ _____

33. $12 + 2.1w = 1.3w$ _____

34. $3.5(j + 4) = 1.4(16 + j)$ _____

35. $4.5 - 1.9m = 20.1 - 2m$ _____

36. $x - 0.09 = 2.22 - 0.1x$ _____

37. $\frac{1}{2}x + 7 = \frac{3}{4}x - 4$ _____

38. $\frac{1}{4}y = \frac{2}{5}y - 1$ _____

39. $\frac{1}{3}z = 3z - \frac{4}{5}$ _____

40. $\frac{a}{2} - \frac{1}{3} = \frac{a}{3} - \frac{1}{2}$ _____

41. $2\left(\frac{1}{3}w + \frac{1}{4}\right) = 4 + \frac{1}{3}w$ _____

42. $\frac{1}{4}(7 + 3r) = -\frac{1}{8}r$ _____

Practice
7.3 Algebraic Applications

1. Mark is a plumber. He charges $28 per job plus $20 per hour. If the total labor bill for a job was $58, how many hours did Mark work? _____

2. Marta is a carpenter. She charges $500 for materials plus $19 per hour for a certain job. How many hours did Marta work on that job if she was paid $1070? _____

3. Fran works in a children's clothing store. She earns $175 per week plus 3.5% of her sales. What must her sales be in order for her to make $525 per week? _____

4. Joe manages a local drug store. His base salary is $300 per week plus 1.8% of the weekly sales at the store. What must the weekly sales be in order for him to make $570 per week? _____

5. The baseball booster club spent $850 to print 1200 baseball programs. They plan to sell each program for $2. How many programs must they sell to make a profit of $500? _____

6. The manager of the local ballet company determined that the cost of costumes, ticket printing, and theater rental for the winter season will be $3600. If each ticket is sold for $12.50, how many tickets must be sold to make a profit of $7000? _____

7. Amy has 2.5 liters of a solution that is 70% acid. She wants to add pure water to make a solution that is 50% acid. How much water should she add? _____

8. Jamil has 36 milliliters of a solution that is 25% salt. How many milliliters of a solution that is 60% salt should he add to make a new solution that is 30% salt? _____

9. A baseball team played 5 more games this season than last season. Last season, the team won 60% of its games, and this season it won 55% of its games. The team won the same number of games this season as last season. Find the number of games played each season. _____

10. In August, Mike's Sporting Goods Store sold 30% of its stock of golf clubs and 35% of its stock of tennis rackets. At the beginning of August, there were 100 more golf clubs than tennis rackets in stock. An equal number of golf clubs and tennis rackets were sold. How many of each item were in stock at the beginning of August? _____

11. The drama club is selling candles. They pay $22 for the advertising brochures and $3.25 for each candle. If they sell each candle for $6.50, how many candles must they sell to break even? _____

Practice
7.4 Geometric Applications

Angles 1 and 2 are complementary. Find x from the values given in Exercises 1–5.

1. $m\angle 1 = x + 3$; $m\angle 2 = 3x - 1$ _____

2. $m\angle 1 = 5x - 2$; $m\angle 2 = 3x$ _____

3. $m\angle 1 = 10x + 12$; $m\angle 2 = 5x + 3$ _____

4. $m\angle 1 = 6x + 1$; $m\angle 2 = 4x - 11$ _____

5. $m\angle 1 = \frac{x}{3} + 2$; $m\angle 2 = x - 10$ _____

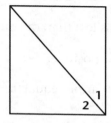

Find the measures of the angles of triangle *MNP*.

6. $m\angle M = 3x + 16$; $m\angle N = 5x$; $m\angle P = 2x - 6$

_____

7. $m\angle M = 15x$; $m\angle N = 12x + 6$; $m\angle P = 7x + 4$

_____

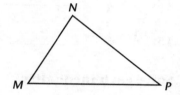

8. $m\angle M = 3x + 10$; $m\angle N = 5x - 2$; $m\angle P = 4x - 8$

_____

9. $m\angle M = \frac{x}{2} + 10$; $m\angle N = x + 30$; $m\angle P = \frac{x}{4}$

_____

Using the figure at the right, find x and the measures of the indicated angles from the information given in Exercises 10–13.

10. $m\angle 1 = 3x - 2$; $m\angle 3 = 4x - 10$

_____

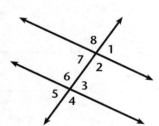

11. $m\angle 6 = 6x + 2$; $m\angle 7 = 4x + 8$

_____

12. $m\angle 7 = 4x - 6$; $m\angle 8 = 5x - 3$

_____

13. $m\angle 1 = 6x - 5$; $m\angle 5 = 3x + 4$

_____

Practice

7.5 Exploring Related Inequalities

Write an inequality that corresponds to each statement.

1. x is less than y. _____

2. W is greater than B. _____

3. a is less than or equal to 10. _____

4. x is greater than or equal to 100. _____

5. r is positive. _____

6. q is non-negative. _____

7. M cannot equal 0. _____

8. V is between 4.5 and 4.6 inclusive. _____

Tell whether each statement is true or false.

9. $6.9 \geq 6.9$ _____

10. $9.66 > 9.606$ _____

11. $10.2 < 10.02$ _____

12. $\frac{1}{2} \leq \frac{1}{3}$ _____

13. $8.91 > 8.901$ _____

14. $0 > -1$ _____

15. $\frac{1}{7} < \frac{1}{5}$ _____

16. $-5 < -2$ _____

17. $-6 \geq -4$ _____

Solve each inequality.

18. $x + 3 < 4$ _____

19. $m - 10 \geq 50$ _____

20. $T - 5 \leq 8$ _____

21. $7 - N \geq 16$ _____

22. $6x < -12$ _____

23. $x + 12 > 10$ _____

24. $3.5b > -7$ _____

25. $2x \geq 6$ _____

26. $\frac{n}{5} > 20$ _____

27. $\frac{r}{-2} \leq 7$ _____

28. $-5t \leq 45$ _____

29. $\frac{x}{4} + 7 < 10$ _____

30. $\frac{-x}{5} + 4 \geq -1$ _____

31. $6x - 2 \leq 4$ _____

32. $x + 3 < 8 - x$ _____

33. $x - 1 < -5$ _____

34. $-5x \geq 2x - 6$ _____

35. $7 - x \leq 3$ _____

36. $16 + \frac{m}{2} < 9$ _____

37. $14 > 4 - \frac{j}{3}$ _____

Practice
7.6 Absolute-Value Equations and Inequalities

**Find the values of *x* that solve each absolute-value equation.
Check your answers.**

1. $|x + 2| = 5$ _____

2. $|x + 6| = 7$ _____

3. $|x - 7| = 4$ _____

4. $|x - 3| = 5$ _____

5. $|4x - 2| = 6$ _____

6. $|3x + 5| = 11$ _____

7. $|-4 + x| = 7$ _____

8. $|x - 2.75| = 0.05$ _____

**Find the values of *x* that solve each absolute-value inequality.
Graph each answer on the number line. Check your answers.**

9. $|x + 2| > 7$

_____

10. $|x + 1| \leq 8$

_____

11. $|-2 - x| \geq 4$

_____

12. $|x + 1| \geq 4$

_____

13. $|x - 3| > 2$

_____

14. $|4 - x| \geq 5$

_____

15. $|x + 2| > 2$

_____

16. $|x - 5| \leq 1$

_____

17. $|x + 2| < 2$

_____

NAME _____ CLASS _____ DATE _____

 Practice
8.1 Representing Linear Functions by Graphs

Graph each list of ordered pairs. State whether they lie on a straight line.

1. $(4, 3), (2, 3), (-2, 3)$

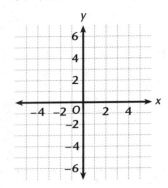

_____

2. $(2, -2), (4, 1), (5, 2)$

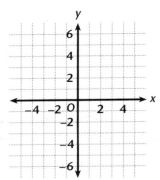

_____

3. $(5, 5), (1, 2), (-3, -1)$

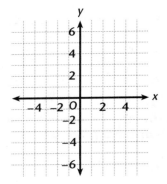

_____

4. $(-4, -2), (-2, 0), (0, 2)$

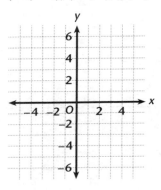

_____

5. $(-3, 4), (0, -1), (3, -5)$

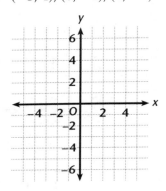

_____

6. $(5, 6), (2, 3), (-1, 0)$

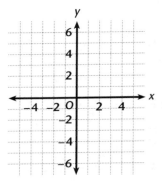

_____

Find the values for y by substituting 1, 2, 3, 4, and 5 for x. Make a table.

7. $y = x + 5$

8. $y = x - 1$

9. $y = 4x$

10. $y = 3x + 2$

Practice
8.2 Exploring Slope

**Examine the graphs below. Which lines have a positive slope?
Which have a negative slope? Which have neither?**

1.

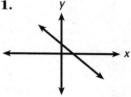

_____

2.

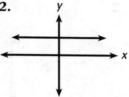

_____

3.

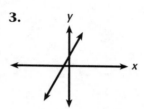

_____

4.

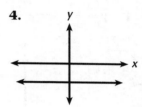

_____

Use the graph to find the slope of each line.

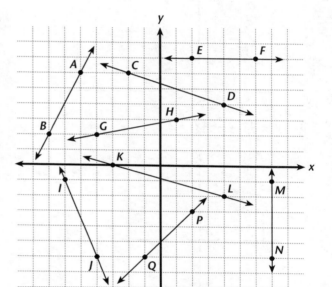

5. $\overleftrightarrow{AB}$ _____   **6.** $\overleftrightarrow{CD}$ _____

7. $\overleftrightarrow{EF}$ _____   **8.** $\overleftrightarrow{GH}$ _____

9. $\overleftrightarrow{IJ}$ _____   **10.** $\overleftrightarrow{KL}$ _____

11. $\overleftrightarrow{MN}$ _____   **12.** $\overleftrightarrow{PQ}$ _____

Find the slope for each line.

13. rise –5, run –5   **14.** rise 2, run 3

_____   _____

15. rise –3, run 4   **16.** rise –2, run –5

_____   _____

Each pair of points is on a line. What is the slope of each line?

17. $A(3, 9), B(1, 5)$ _____   **18.** $A(7, 5), B(2, 4)$ _____

19. $A(-3, 10), B(-5, -4)$ _____   **20.** $A(5, 2), B(2, -1)$ _____

21. $A(3, -2), B(-1, 3)$ _____   **22.** $A(-1, 3), B(5, 3)$ _____

23. $A(1, 8), B(-1, 7)$ _____   **24.** $A(2, 6), B(3, -4)$ _____

25. $A(0, 4), B(3, -2)$ _____   **26.** $A(6, -1), B(5, 6)$ _____

27. $A(-9, 9), B(7, -2)$ _____   **28.** $A(3, 7), B(-1, 0)$ _____

Practice

8.3 Exploring Graphs of Linear Functions

Draw each pair of graphs on the same axes. In each case, tell what is the same and what is different about the graphs.

1. $y = \frac{1}{4}x$; $y = -\frac{1}{4}x$

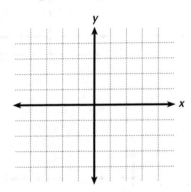

_____

2. $y = 2x - 1$; $y = -2x - 1$

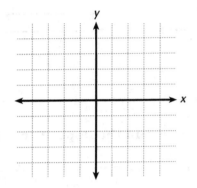

_____

3. $y = 4x$; $y = 4x - 2$

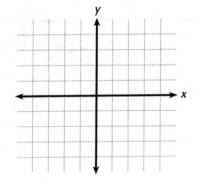

_____

4. $y = \frac{1}{3}x + 4$; $y = \frac{1}{3}x - 1$

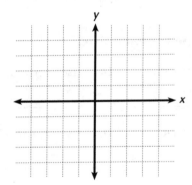

_____

Find the equation of the line that passes through the origin and each of the given points.

5. $(3, 6)$ _____

6. $(2, -1)$ _____

7. $(4, 3)$ _____

8. $(-1, 8)$ _____

9. $(5, 5)$ _____

10. $(7, 2)$ _____

Guess what each line will look like when it is graphed. Use your own graph paper to check your guess by graphing.

11. $y = 6x$ _____

12. $y = -\frac{1}{2}x + 5$ _____

13. $y = -4$ _____

Practice

8.4 The Slope-Intercept Form

Give the coordinates of the point where the line for each equation crosses the *y*-axis.

1. $y = 3x + 4$ _____

2. $y = 2x - 3$ _____

3. $y = \frac{1}{2}x$ _____

4. $y = 2 - x$ _____

Write an equation for the graph of each line.

5.

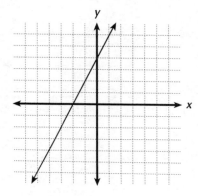

_____

6.

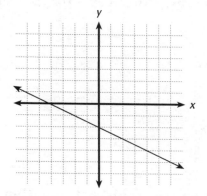

_____

Write the equation for each line.

7. with slope 2 and *y*-intercept 4 _____

8. with slope -3 and *y*-intercept 1 _____

9. through $(0, -4)$ and with slope 2 _____

10. through $(0, 6)$ and with slope $\frac{1}{2}$ _____

11. with slope $-\frac{3}{4}$ and *y*–intercept -3 _____

12. through $(0, 1)$ and with slope 1.5 _____

Write the equation for the line passing through each pair of points.

13. $(3, 8), (2, 6)$ _____

14. $(0, -6), (-3, 3)$ _____

15. $(-2, -4), (5, -1)$ _____

16. $(-1, -2), (-3, -4)$ _____

Practice
8.5 Other Forms for Equations of Lines

Write each equation in standard form.

1. $2x = -5y + 11$ _____

2. $3y = -x - 20$ _____

3. $4x - 7y + 15 = 0$ _____

4. $9x = 6y$ _____

5. $2x + 10 = 3y - 1$ _____

6. $2x = \frac{1}{2}y + 3$ _____

Find the *x*- and *y*- intercepts for each equation.

7. $x + y = 5$ _____

8. $3x + 5y = 15$ _____

9. $4x - 3y = 12$ _____

10. $x - 3y = 6$ _____

11. $x - y = -3$ _____

12. $4x = -5y$ _____

13. $2x + y = 1$ _____

14. $x = \frac{2}{3}y$ _____

15. $\frac{x}{4} - y = 2$ _____

16. $x = -6y - 2$ _____

Use intercepts to graph each equation.

17. $2x - y = -4$

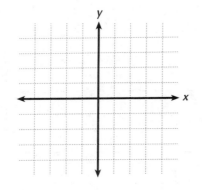

18. $x - 2y = 2$

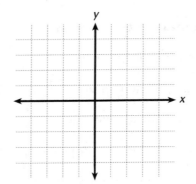

Write the equation of each line.

19. through (4, 5) and with slope 1 _____

20. crosses the *x*-axis at $x = 3$ and the *y*-axis at $y = 6$ _____

21. through (1, 6) and with slope 2 _____

22. through (3, 7) and (0, −2) _____

23. through (1, 5) and (−3, 1) _____

Practice
8.6 Vertical and Horizontal Lines

Find the slope for each equation. Using your own graph paper, plot points, graph the line for each equation, and indicate if the line is vertical, horizontal, or neither.

1. $x = 3$ _____

2. $y = -5$ _____

3. $y = 2x$ _____

4. $y = 5$ _____

5. $x = -1$ _____

6. $y = -\frac{1}{3}x$ _____

Match each equation with the appropriate description.

_____ **7.** $y = 5x$ **a.** a line through the origin with slope $\frac{1}{5}$

_____ **8.** $x = 5$ **b.** a line with slope -1 and y-intercept 5

_____ **9.** $y = 5x^2$ **c.** a horizontal line 5 units above the origin

_____ **10.** $x = 5y$ **d.** a line through the origin with slope 5

_____ **11.** $y = 5$ **e.** something other than a straight line

_____ **12.** $x + y = 5$ **f.** a vertical line 5 units to the right of the origin

Write two equivalent forms of each equation or "undefined slope," when appropriate.

Given	Slope-intercept form	Standard form
13. $y = 3x$		
14. $y = 5$		
15. $x + y - 2 = 0$		
16. $x = -6$		
17. $x = 7y$		

Determine whether each line is vertical or horizontal.

18. $y = 7$ _____

19. $x = 6$ _____

20. $y = -14$ _____

21. $y = 0$ _____

22. $x = -3$ _____

23. $x = 20$ _____

Practice
8.7 Parallel and Perpendicular Lines

Write the slope of a line that is parallel to each line.

1. $y = 2x - 5$ _____

2. $y = -x + 2$ _____

3. $3x + y = 10$ _____

4. $5x - y = 11$ _____

5. $x + 2y = 6$ _____

6. $2x - 3y = 9$ _____

7. $4x + y = 3$ _____

8. $x + 2y = 14$ _____

Write the slope of a line that is perpendicular to each line.

9. $y = 4x + 6$ _____

10. $y = -\frac{1}{5}x - 3$ _____

11. $x + y = 7$ _____

12. $6x - y = 14$ _____

13. $x + 7y = -21$ _____

14. $5x - 4y = 12$ _____

15. $y = \frac{1}{3}x + 2$ _____

16. $2y = -2x - 8$ _____

Write an equation for the line containing the point (6, −2) and

17. parallel to the line $2x + y = 5$. _____

18. perpendicular to the line $y = -3x + 4$. _____

Write an equation for the line containing the point (−6, 5) and

19. parallel to the line $x + 2y = 6$. _____

20. perpendicular to the line $3x - 4y = -8$. _____

Write an equation for the line containing the point (−3, 2) and

21. parallel to the line $y = -4$. _____

22. perpendicular to the line $y = -4$. _____

Write an equation for the line containing the point (−1, 2) and

23. parallel to the line $y = x - 6$. _____

24. perpendicular to the line $y = -x$. _____

Practice
8.8 Exploring Linear Inequalities

Graph each inequality.

1. $y < 2x$

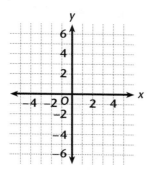

2. $y \geq x + 1$

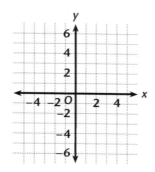

3. $y \leq 3x - 2$

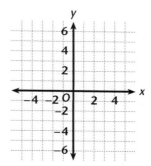

4. $2y > 6x$

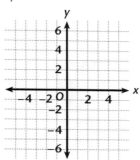

5. $x > -2$

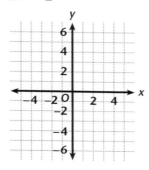

6. $y \leq 5$

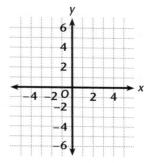

7. $y \leq \frac{1}{3}x - 1$

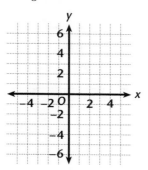

8. $y > -x - 4$

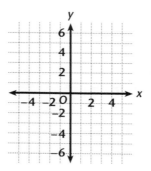

9. $-2x + 2y - 4 < 0$

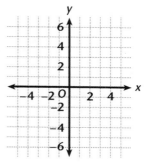

Write an inequality for each graph.

10.

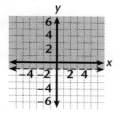

_____

11.

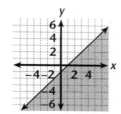

_____

12.

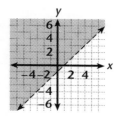

_____

Practice

9.1 Solving Systems by Graphing

Estimate the solution to each system of equations.

1.

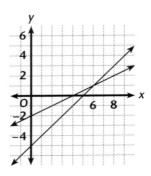

_____

2.

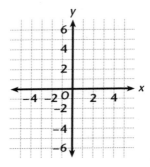

_____

3.

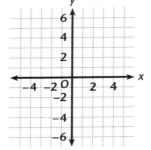

_____

**Solve each system of equations by graphing on the grid provided.
If necessary, give an approximate solution.**

4. $\begin{cases} x + y = 1 \\ 2x + y = 4 \end{cases}$

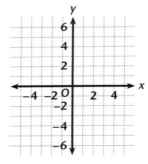

_____

5. $\begin{cases} 3x + y = 1 \\ x = -1 \end{cases}$

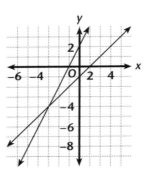

_____

6. $\begin{cases} y = \frac{1}{2}x + 1 \\ y = x \end{cases}$

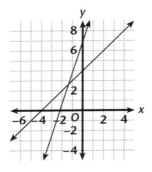

_____

7. $\begin{cases} 4x - 2y = -1 \\ y - x = 2 \end{cases}$

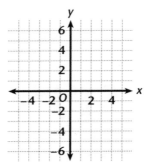

_____

8. $\begin{cases} x = -3 \\ y = -2 \end{cases}$

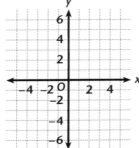

_____

9. $\begin{cases} y = 4x - 3 \\ y = 1 \end{cases}$

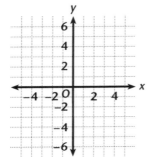

_____

Practice
9.2 The Substitution Method

Use the substitution method to solve each system of equations. Check your answer by substituting your solution in the original equations.

1. $\begin{cases} y = 3x \\ x + y = 4 \end{cases}$ _____

2. $\begin{cases} m = 5n \\ 2m + 5n = 15 \end{cases}$ _____

3. $\begin{cases} y = 8 \\ 2y - x = 22 \end{cases}$ _____

4. $\begin{cases} m - 4n = 12 \\ 2m - 10 = n \end{cases}$ _____

5. $\begin{cases} x = -3 \\ 3x + 2y = -8 \end{cases}$ _____

6. $\begin{cases} x = 12 - y \\ x - 2y = -8 \end{cases}$ _____

7. $\begin{cases} x = y + 1 \\ x + y = 3 \end{cases}$ _____

8. $\begin{cases} y = 3x - 4 \\ 2x - 3y = -9 \end{cases}$ _____

9. $\begin{cases} 2p - q = 7 \\ 3p - 4q = 8 \end{cases}$ _____

10. $\begin{cases} 3m - n = 4 \\ 6m + 2n = 8 \end{cases}$ _____

11. $\begin{cases} r = 8 - 4t \\ 2r - 5t = 29 \end{cases}$ _____

12. $\begin{cases} y = -4x \\ x = 2y - 7 \end{cases}$ _____

13. $\begin{cases} x + y = 2 \\ 2x - y = 10 \end{cases}$ _____

14. $\begin{cases} 3x - 4y = 11 \\ x - 2y = 10 \end{cases}$ _____

15. $\begin{cases} a - b = 1 \\ a + b = 11 \end{cases}$ _____

16. $\begin{cases} z + 7w = -14 \\ 6z + 7w = 21 \end{cases}$ _____

17. $\begin{cases} 2c - 3d = -24 \\ c + 6d = 18 \end{cases}$ _____

18. $\begin{cases} p + 14q = 84 \\ 2p - 7q = -7 \end{cases}$ _____

19. $\begin{cases} x - 2y = 6 \\ \frac{1}{2}x + 2y = 12 \end{cases}$ _____

20. $\begin{cases} y = -\frac{2}{5}x \\ y + \frac{4}{5}x = 2 \end{cases}$ _____

21. $\begin{cases} x + y = 70 \\ 0.1x + 0.3y = 15 \end{cases}$ _____

22. $\begin{cases} y = 1.2x - 5 \\ y = -1.3x + 30 \end{cases}$ _____

23. $\begin{cases} 0.5m + 5n = 19 \\ m + 10 = 2n \end{cases}$ _____

24. $\begin{cases} \frac{3}{4}z - w = -4 \\ 4w = 2z + 5 \end{cases}$ _____

Practice

9.3 Exploring Elimination by Addition

Model each system of equations with algebra tiles. Then solve the system by using elimination by addition.

1. $\begin{cases} x + y = 4 \\ x - y = 8 \end{cases}$ _____

2. $\begin{cases} 2x - y = -5 \\ 2x + 3y = 3 \end{cases}$ _____

3. $\begin{cases} x - 2y = 5 \\ 5x - 2y = 9 \end{cases}$ _____

4. $\begin{cases} x - 4y = 8 \\ x + 3y = 1 \end{cases}$ _____

Solve each system of equations by using elimination by addition.

5. $\begin{cases} x + y = 12 \\ x - y = 4 \end{cases}$ _____

6. $\begin{cases} x - y = 6 \\ x + y = 2 \end{cases}$ _____

7. $\begin{cases} 2m + n = 3 \\ 3m - n = 2 \end{cases}$ _____

8. $\begin{cases} y - 2x = -4 \\ 3y + 2x = 4 \end{cases}$ _____

9. $\begin{cases} 3x - 2y = 8 \\ 8x + 2y = 14 \end{cases}$ _____

10. $\begin{cases} x + y = -1 \\ x - 5y = -7 \end{cases}$ _____

11. $\begin{cases} y - 3x = 3 \\ 2y + 3x = -12 \end{cases}$ _____

12. $\begin{cases} 5x - y = -6 \\ x - y = -2 \end{cases}$ _____

13. $\begin{cases} 2x + 2y = 14 \\ 2x + 3y = 6 \end{cases}$ _____

14. $\begin{cases} m - 3n = -2 \\ m + 3n = 6 \end{cases}$ _____

15. $\begin{cases} 4x - 5y = -2 \\ 2x = -5y - 16 \end{cases}$ _____

16. $\begin{cases} y = 3 - x \\ 3x + y = 1 \end{cases}$ _____

17. $\begin{cases} \frac{1}{3}x + y = -2 \\ \frac{1}{3}x - y = -\frac{2}{3} \end{cases}$ _____

18. $\begin{cases} \frac{2}{5}x = -y \\ \frac{4}{5}x = y - 2 \end{cases}$ _____

19. $\begin{cases} 1.5x - 0.5y = 3.5 \\ 3x + 0.5y = 1 \end{cases}$ _____

20. $\begin{cases} 0.2x + 0.2y = -1.4 \\ -0.2x - 0.4y = 2 \end{cases}$ _____

21. $\begin{cases} x - \frac{1}{3}y = \frac{7}{3} \\ x + \frac{1}{6}y = \frac{1}{3} \end{cases}$ _____

22. $\begin{cases} 0.3m + 1.9n = -3.8 \\ 0.1m - 1.9n = 11.4 \end{cases}$ _____

Practice

9.4 Elimination by Multiplication

Solve each system of equations by using elimination by multiplication. Check your solution either by graphing or by substitution.

1. $\begin{cases} 2x + 5y = 3 \\ x - 3y = 7 \end{cases}$ _____

2. $\begin{cases} m + 3n = 13 \\ 3m + 3n = 9 \end{cases}$ _____

3. $\begin{cases} 2x + y = 3 \\ 4x - 4y = 0 \end{cases}$ _____

4. $\begin{cases} 2a + 3b = 9 \\ 3a + 2b = 12 \end{cases}$ _____

5. $\begin{cases} 5r - 7s = -10 \\ r - 4s = 11 \end{cases}$ _____

6. $\begin{cases} y + 5x = 12 \\ 3y - 2x = 19 \end{cases}$ _____

7. $\begin{cases} 2c - 5d = 1 \\ 3c - 7d = 3 \end{cases}$ _____

8. $\begin{cases} 2x + 3y = 4 \\ 3x - 5y = 9 \end{cases}$ _____

9. $\begin{cases} 2x + 3y = 7 \\ 5x + 4y = 14 \end{cases}$ _____

10. $\begin{cases} z = 12 - 5w \\ 2w = 3z - 19 \end{cases}$ _____

11. $\begin{cases} 2y - 3x = 0 \\ x + y = -10 \end{cases}$ _____

12. $\begin{cases} 2x - 5y = 1 \\ 4x - 3y = 9 \end{cases}$ _____

13. $\begin{cases} 8y = 5x \\ 2x - 3y = -17 \end{cases}$ _____

14. $\begin{cases} 2m + 3n = 13 \\ 6m - 5n = -3 \end{cases}$ _____

15. $\begin{cases} -3x + 4y = 20 \\ 2.5x - 2y = -6 \end{cases}$ _____

16. $\begin{cases} 0.5x + 0.5y = -1 \\ 0.3x + 0.6y = -0.6 \end{cases}$ _____

17. $\begin{cases} \frac{1}{2}m - \frac{1}{3}n = 1 \\ 2m + \frac{2}{3}n = -8 \end{cases}$ _____

18. $\begin{cases} -2x + 3y = 10 \\ 2x - \frac{3}{2}y = 4 \end{cases}$ _____

19. $\begin{cases} \frac{1}{4}x + \frac{1}{4}y = \frac{7}{4} \\ x + \frac{3}{2}y = 3 \end{cases}$ _____

20. $\begin{cases} 3x - y = 7 \\ 3x + \frac{1}{2}y = 1 \end{cases}$ _____

21. $\begin{cases} y - x = -3 \\ \frac{1}{2}x + \frac{1}{3}y = 2 \end{cases}$ _____

22. $\begin{cases} y + x = 1 \\ \frac{1}{2}y - \frac{1}{3}x = 1 \end{cases}$ _____

Practice

9.5 Systems of Linear Inequalities

Circle the letter of the points that are solutions to each system of inequalities.

1. $\begin{cases} x - y \geq 2 \\ \quad y \leq 2 \end{cases}$ **a.** $(1, 2)$ **b.** $(1, -2)$ **c.** $(-1, 2)$

2. $\begin{cases} 1 + y \leq 2x \\ \quad y + x \geq 6 \end{cases}$ **a.** $(2, 4)$ **b.** $(1, -1)$ **c.** $(3, 5)$

3. $\begin{cases} y \leq 2x - 5 \\ 4x + 3y \geq 2 \end{cases}$ **a.** $(2, 4)$ **b.** $(6, -1)$ **c.** $(1, -3)$

Write each system of inequalities graphed.

4.

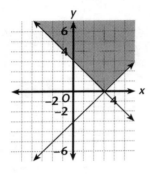

5.

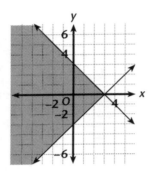

Graph each system of inequalities.

6. $\begin{cases} \quad y > 2 \\ y < x + 4 \end{cases}$

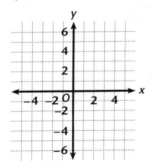

7. $\begin{cases} x \leq 2 \\ y \geq -2 \end{cases}$

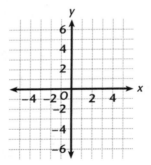

8. $\begin{cases} 2x + y \leq 1 \\ 3x - y > 2 \end{cases}$

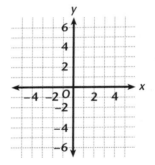

9. $\begin{cases} \quad x \geq 4 \\ y \geq \frac{1}{2}x - 1 \end{cases}$

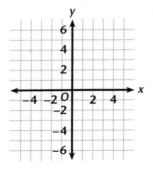

10. $\begin{cases} y + 3 \geq x \\ 4x - y > 2 \end{cases}$

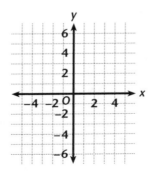

11. $\begin{cases} \quad y > -3 \\ x - 2y \geq 3 \\ \quad x + y < 5 \end{cases}$

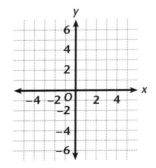

Practice

10.1 Previewing Exponential Functions

What are the next three terms of each sequence? State whether the sequence shows linear or exponential growth.

1. 4, 7, 10, 13, 16, . . . _____

2. 1, 2, 4, 8, 16, . . . _____

3. 11, 22, 33, 44, 55, . . . _____

4. 25, 5, 1, $\frac{1}{5}$, $\frac{1}{25}$, . . . _____

5. 1, 3, 9, 27, 81, . . . _____

6. 3, 6, 12, 24, 48, . . . _____

7. Write the first five terms of a sequence that starts with 2 and triples the previous number.

   _____

8. Write the first five terms of a sequence that starts with 4 and doubles the previous number.

   _____

9. Write the first five terms of a sequence that starts with 500 and halves the previous number.

   _____

10. Generate a table for $y = 4^x$.

x	1	2	3	4	5
y					

11. Generate a table for $y = 5^x$.

x	1	2	3	4	5
y					

Suppose that you are standing 16 feet from a wall. Each minute you walk one-half of the distance to the wall. How far from the wall will you be after

12. 1 minute? _____

13. 2 minutes? _____

14. 4 minutes? _____

15. 6 minutes? _____

Practice
10.2 Exploring Quadratic Functions

Match each table with the correct type of relationship.

A. linear   **B.** exponential   **C.** quadratic   **D.** none of these

_____ **1.**

x	1	2	3	4	5
y	3	5	7	9	11

_____ **2.**

x	1	2	3	4	5
y	2	8	18	32	50

_____ **3.**

x	1	2	3	4	5
y	1	3	8	14	17

_____ **4.**

x	1	2	3	4	5
y	0	3	8	15	24

_____ **5.**

x	1	2	3	4	5
y	0	2	6	12	20

_____ **6.**

x	1	2	3	4	5
y	2	4	8	16	32

_____ **7.**

x	1	2	3	4	5
y	6	3	2	3	6

_____ **8.**

x	1	2	3	4	5
y	5	10	13	14	13

Match each relationship with the correct graph.

A. linear   **B.** exponential   **C.** quadratic   **D.** none of these

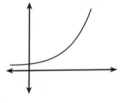

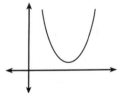

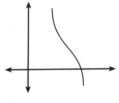

9. _____   **10.** _____   **11.** _____   **12.** _____

What happens to the area of a square if you

13. halve the length of each side? _____

14. multiply the length of each side by 5? _____

15. divide the length of each side by 5? _____

 **Practice**

10.3 Previewing Reciprocal Functions

Find the reciprocal of each number.

1. 6 _____

2. 10 _____

3. $\frac{16}{5}$ _____

4. $\frac{8}{9}$ _____

5. Complete the table. Let the *y*-value for each *x*-value be the reciprocal of the *x*-value. Graph the function on the grid.

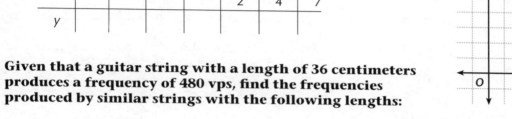

x	1	2	4	7	$\frac{1}{2}$	$\frac{1}{4}$	$\frac{1}{7}$
y							

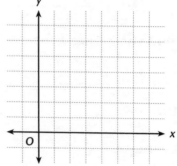

Given that a guitar string with a length of 36 centimeters produces a frequency of 480 vps, find the frequencies produced by similar strings with the following lengths:

6. 72 cm _____

7. 18 cm _____

8. 27 cm _____

Plot several ordered pairs for each of the given functions. Join the points with a smooth curve.

9. $y = \frac{5}{x}$

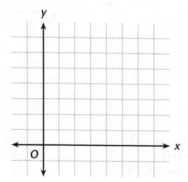

10. $y = \frac{10}{x}$

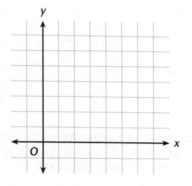

The Environmental Club wants to raise $200. How much will each person have to contribute if there are

11. 20 people? _____

12. 50 people? _____

13. 500 people? _____

How fast do you have to travel to make a 360-mile trip in

14. 1 hour? _____

15. 6 hours? _____

16. 4 hours? _____

17. 8 hours? _____

18. $\frac{1}{2}$ hour? _____

19. 15 minutes? _____

Practice

10.4 Previewing Other Functions

Find the absolute value (ABS).

1. 13 _____ **2.** −21 _____ **3.** −4.2 _____

4. 18.7 _____ **5.** −1.38 _____ **6.** 0 _____

Find the greatest integer (INT).

7. $\frac{5}{2}$ _____ **8.** 3.7 _____ **9.** 10.6 _____

10. 48 _____ **11.** 17.9 _____ **12.** $4\frac{1}{4}$ _____

Evaluate.

13. $|31|$ _____ **14.** $|-9.6|$ _____ **15.** $|0.8|$ _____

16. ABS(−4.3) _____ **17.** ABS$\left(-\frac{1}{2}\right)$ _____ **18.** ABS(9) _____

19. INT(4.3) _____ **20.** INT(10.63) _____ **21.** INT$\left(\frac{1}{2}\right)$ _____

22. $|6|$ _____ **23.** $|-6|$ _____ **24.** $-|6|$ _____

25. $-|-6|$ _____ **26.** $|2.3|$ _____ **27.** $|-2.3|$ _____

28. $-|2.3|$ _____ **29.** $-|-2.3|$ _____ **30.** $-(-|-2.3|)$ _____

31. INT(−6.7) _____ **32.** INT$\left(-12\frac{1}{3}\right)$ _____ **33.** INT(0.45) _____

Luanne has a large jar of pennies and is placing them in coin wrappers. If each wrapper holds 50 pennies, how many wrappers can she fill with the following number of pennies?

34. 71 _____ **35.** 268 _____ **36.** 499 _____

The Chess Club is selling homemade cookies to raise money. If the cookies are packed 6 to a bag, how many full bags can be packed from the following number of cookies?

37. 57 _____ **38.** 181 _____ **39.** 221 _____

 **Practice**

10.5 Identifying Types of Functions

Match each graph with the correct type of function.

| **A.** linear | **B.** exponential | **C.** quadratic |
| **D.** reciprocal | **E.** absolute-value | **F.** integer |

_____ **1.**

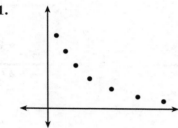

_____ **2.**

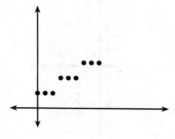

_____ **3.**

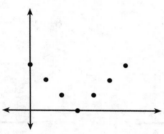

_____ **4.**

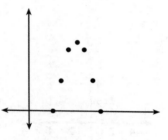

_____ **5.**

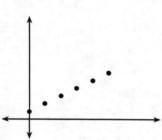

_____ **6.**

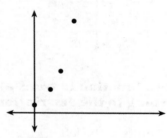

Use the method of differences to determine whether the relationship is linear, quadratic, or neither.

7.

x	1	2	3	4	5
y	1	2	4	12	25

_____

8.

x	1	2	3	4	5
y	0	5	12	21	32

_____

9.

x	1	2	3	4	5
y	3	8	13	18	23

_____

10.

x	1	2	3	4	5
y	6	12	20	30	42

_____

11.

x	0	1	2	3	4
y	1	2	4	8	16

_____

12.

x	0	1	2	3	4
y	2	3	4	5	6

_____

Practice
10.6 Exploring Transformations

Sketch a graph of each of the following functions on the grid provided:

1. $y = |x| + 3$

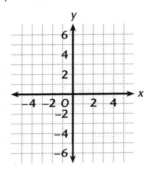

2. $y = |x - 3|$

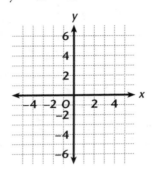

3. $y = |x| - 1$

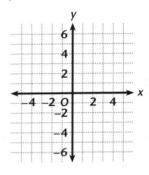

4. $y = (x - 2)^2$

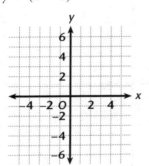

5. $y = (x + 1)^2 - 1$

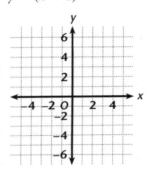

6. $y = -|x - 3|$

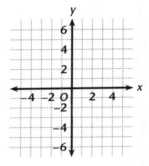

Write a function to translate the parent function $y = |x|$ according to the descriptions below.

7. 5 units to the left _____

8. 4 units down _____

9. 2.5 units to the right _____

10. 3 units up _____

11. 4 units up and 2 units to the left _____

12. 16 units down and 5 units to the right _____

Write a function to translate the parent function $y = x^2$ according to the descriptions below.

13. 6 units to the right _____

14. 1 unit down _____

15. 25 units up _____

16. 6 units to the left _____

17. 2 units down and 4 units to the right _____

18. 10 units up and 3 units to the left _____

Practice
11.1 Exploring Graphs

The graphs below show the monthly CD sales for two different music stores.

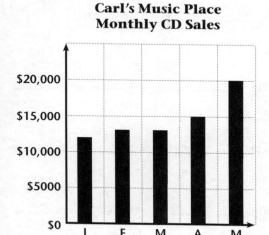

1. During which month were the sales at Carl's Music Place the greatest?

What were the sales? _____

2. During which month were the sales at Carl's Music Place the least?

What were the sales? _____

3. During which month were the sales at The Music Store the greatest?

What were the sales? _____

4. During which month were the sales at The Music Store the least?

What were the sales? _____

5. What were the CD sales for Carl's Music Place for January through

May? _____

6. What were the CD sales for The Music Store for January through May? _____

7. Which store appears to have a longer bar to represent April sales?

Which company actually had greater sales in April? _____

8. Describe how displaying the graphs together can be misleading. _____

_____

Practice
11.2 Exploring Measures of Central Tendency

Find the mean, median, mode, and range for each data set.

1. 13, 13, 10, 8, 7, 6, 4, 5

　　mean _____　　median _____　　mode _____　　range _____

2. 20, 30, 35, 24, 36, 47, 48

　　mean _____　　median _____　　mode _____　　range _____

3. 2, 5, 4, 1, 6, 7, 4, 3, 2, 1

　　mean _____　　median _____　　mode _____　　range _____

4. 130, 140, 135, 125, 160, 175

　　mean _____　　median _____　　mode _____　　range _____

5. 16, 18, 39, 200, 31, 39

　　mean _____　　median _____　　mode _____　　range _____

The Sleep Shop conducted a survey to determine the average number of hours that people sleep at night. The results are shown at the right. Use this data for Exercises 6–12.

Number of Hours Spent Sleeping at Night				
5	8	6	7	4
9	8	7	5	9
8	10	7	7	8
6	8	8	7	8
9	8	7	5	9
10	7	8	8	6

6. Make a frequency table for the data.

Find the measures of central tendency for the data.

7. mean _____　　**8.** median _____　　**9.** mode _____　　**10.** range _____

11. Which measure of central tendency do you think gives the best indication of the number of hours the "typical" person spends

sleeping each night? Explain. _____

_____

12. Suppose another person was surveyed who said that he spends 3 hours sleeping at night. How would this affect the mean, median,

mode, and range? _____

_____

Practice
11.3 Graphing Data

In the list at right are the ages of the first 42 presidents of the United States when they were sworn into office.

57	61	57	57	58	57
61	54	68	51	49	64
50	48	65	52	56	46
54	49	50	47	55	55
54	42	51	56	55	51
54	51	60	62	43	55
56	61	52	69	64	46

1. Use the data to make a stem-and-leaf plot.

2. What is the range of the data? _____

3. What is the median of the data? _____

4. What are the lower and upper quartiles for this data? _____

5. What is the mean of the data? _____

6. What is the mode of the data? _____

7. What is the average age of a president of the United States when he is sworn into office? What measure of central tendency do you think best

answers this question? Why? _____

_____

8. Construct a box-and-whisker plot for this data.

Practice
11.4 Circle Graphs

Jason made a circle graph to show how he spends his time each weekday. Use the graph to answer Exercises 1–5.

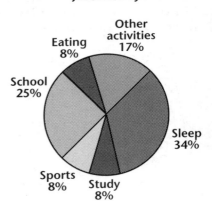

Jason's Day

1. How much time does he spend sleeping?

   _____

2. How much time does he spend at school each day?

   _____

3. How much time does he spend participating in sports? At which other activities does he spend an equal amount of time?

   _____

4. Which individual categories account for less than one-fourth of Jason's time?

   _____

5. Which individual categories account for more than one-fourth of Jason's time?

   _____

The following table shows the budget for publishing one edition of a neighborhood newspaper.

	Printing	Editorial	Design	Photos
Cost	$2000	$537.10	$351.90	$911
Percent				

6. Find the total amount spent to publish one edition of the newspaper.

   _____

7. Complete the table by finding the percent of the budget that is spent in each category.

8. Use the information from the table to make a circle graph that shows the percent of the budget spent in each category.

Practice

11.5 Exploring Scatter Plots and Correlation

Describe the correlation as strong positive, strong negative, or little to none. Explain the reason for your answer.

1.

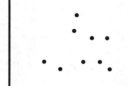

_____

_____

_____

_____

2.

_____

_____

_____

_____

3.

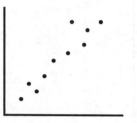

_____

_____

_____

_____

The chart shows the average time that a person can survive in water at a particular temperature.

Water temperature (°F)	37°	45°	55°	60°
Average survival time (min)	7	18	29	60

4. Use the grid at the right to make a scatter plot of water temperature against average survival time.

5. Describe the correlation between temperature and survival time. Explain your reasoning.

_____

_____

_____

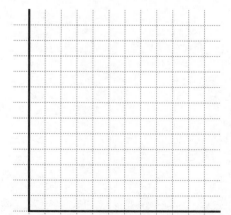

 **Practice**

11.6 Finding Lines of Best Fit

Tell whether the correlation for the scatter plot is positive, negative, or neither.

1.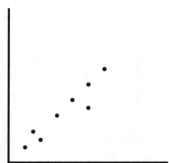

_____

2.

_____

3.

_____

4. For which scatter plot(s) above does it seem reasonable to fit a straight line to the data? _____

5. Which scatter plot above would most likely have a correlation of −0.85? _____

6. Which scatter plot would most likely have a correlation of 0.95? _____

7. Use the grid to plot points (0, 1), (1, 2), (2, 4), and (3, 8). Then use a straightedge to estimate the line of best fit.

8. Is the correlation nearest to −1, 0, or 1?

_____

9. What shape do the points suggest?

_____

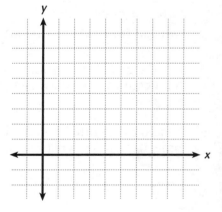

Practice
12.1 Exploring Circles

Find the circumference and area, to the nearest hundredth, of each circle with the given radius or diameter.

1. radius of 6 feet _____

2. radius of 2.5 centimeters _____

3. radius of 13 inches _____

4. radius of 7 meters _____

5. diameter of 8 yards _____

6. diameter of 16 centimeters _____

7. radius of $9\frac{1}{4}$ feet _____

8. diameter of 9.5 meters _____

Complete the table. Round answers to the nearest hundredth.

	Radius	Diameter	Area	Circumference
9.		7 cm		
10.		11 yd		
11.	9 ft			
12.	15 cm			
13.	3.4 ft			
14.	1.6 m			
15.		20 in.		
16.				12 in.
17.			100 m²	
18.			25 ft²	

Practice
12.2 Exploring Surface Area and Volume

Find the surface area and volume of each cube with the given edge length.

1. 3 meters _____

2. 4 inches _____

3. 6 feet _____

4. 8 centimeters _____

5. 1.5 inches _____

6. 6.25 meters _____

Find the surface area and volume of each rectangular solid with the indicated dimensions.

7. 4 m × 5 m × 5 m _____

8. 2 in. × 3 in. × 4 in. _____

9. 1 ft × 4 ft × 6 ft _____

10. 8 m × 8 m × 9 m _____

11. 1.2 m × 1.3 m × 2 m _____

12. 5 cm × 10 cm × 15 cm _____

Complete the table for rectangular solids. Round answers to the nearest hundredth.

	Length	Width	Height	Surface area	Volume
13.	5 cm	6 cm	6 cm		
14.	3 in.	8 in.	9 in.		
15.	4 ft	6 ft	10 ft		
16.	1.5 m	3 m	5 m		
17.	$\frac{1}{2}$ ft	3 ft	$5\frac{1}{2}$ ft		
18.	8.55 m	9.45 m	10.35 m		
19.	6 m	7 m			336 m^3
20.	3 in.	10 in.			600 in.^{3}

Practice
12.3 Prisms

1. Find the volume of a right triangular prism with a base area of 10 square inches and a height of 6 inches.

   _____

2. Find the lateral surface area of a right triangular prism with a base area of 24 square centimeters, a base perimeter of 24 centimeters, and a height of 10 centimeters.

   _____

3. Find the total surface area of a right triangular prism with a base area of 30 square feet, a base perimeter of 30 feet, and a height of 8 feet.

   _____

4. Find the lateral surface area of a right rectangular prism with a base perimeter of 42 inches and a height of 17 inches.

   _____

5. Find the total surface area of a right rectangular prism with a base area of 18 square inches, a base perimeter of 18 inches, and a height of 8 inches.

   _____

6. Find the volume of a right rectangular prism for which the base dimensions are 8.5 centimeters by 3.4 centimeters and the height is 14 centimeters.

   _____

7. Find the volume of a right hexagonal prism with a base area of 21.25 square meters and a height of 20.5 meters.

   _____

8. Find the volume of a right pentagonal prism with a base area of 9 square feet and a height of 12 feet.

   _____

9. Find the total surface area of a right octagonal prism with a base area of 50 square feet, a base perimeter of 26 feet, and a height of 20 feet.

   _____

10. Find the total surface area of a right rectangular prism with a base area of 25 square feet, a base perimeter of 20 feet, and a height of 5 feet.

    _____

Practice
12.4 Cylinders

Find the surface area and volume of each right cylinder described. Round answers to the nearest hundredth.

1. 4-in. radius and 8-in. height _____

2. 15-ft radius and 10-ft height _____

3. 8-m radius and 12-m height _____

4. 50-cm radius and 90-cm height _____

5. 6-ft diameter and 9-ft height _____

6. 2.5-cm radius and 3.5-cm height _____

7. 10-m diameter and 15-m height _____

8. 25-ft diameter and 12-ft height _____

9. 16-in. height and 6-in. radius _____

10. 5.4-m height and 5.8-m diameter _____

11. 105-ft height and 55-ft diameter _____

12. 10.7-cm height and 15.9-cm radius _____

13. 9-yd height and 13-yd diameter _____

14. 4.7-ft height and 12.7-ft diameter _____

15. 57-ft height and 65-ft diameter _____

16. 14-m height and 2.5-m diameter _____

Practice

12.5 Volume of Cones and Pyramids

Find the volume of each right cone or pyramid. Round answers to the nearest hundredth.

1. cone with a radius of 5 inches and a height of 16 inches _____

2. cone with a radius of 8 centimeters and a height of 12 centimeters _____

3. cone with a diameter of 12 feet and height of 9 feet _____

4. cone with a diameter of 10 meters and a height of 11 meters _____

5. cone with a diameter of 15 feet and a height of 12 feet _____

6. cone with a diameter of 5.5 meters and a height of 7 meters _____

7. cone with a radius of $7\frac{1}{4}$ inches and a height of 9 inches _____

8. pyramid with a base area of 12 square inches and a height of 10 inches _____

9. pyramid with a base area of 15 square meters and a height of 18 meters _____

10. pyramid with a base area of 20 square feet and a height of 9 feet _____

11. pyramid with a base area of 24 square centimeters and a height of 8 centimeters _____

12. pyramid with a base area of 15 square inches and a height of $8\frac{3}{4}$ inches _____

13. pyramid with a base area of 35 square yards and a height of $17\frac{1}{2}$ yards _____

14. pyramid with a base area of 18 square meters and a height of 17.5 meters _____

15. pyramid with a base area of 7 square inches and a height of 8.5 inches _____

Practice

12.6 Surface Area of Cones and Pyramids

Find the lateral area and surface area of each right cone described. Round answers to the nearest hundredth.

1. a slant height of 5 meters and a diameter of 10 meters _____

2. a slant height of 7 yards and a diameter of 15 yards _____

3. a slant height of 6 feet and a radius of 12 feet _____

4. a slant height of 12 yards and a radius of 8 yards _____

5. a slant height of 5.6 centimeters and a diameter of 12.8 centimeters _____

Find the lateral area and surface area of each right regular pyramid described.

6. a square base with side lengths of 5 centimeters and a slant height of

   6 centimeters _____

7. a square base with side lengths of 8 inches and a slant height of

   12 inches _____

Find the surface area of each right cone described. Round answers to the nearest hundredth.

8. a radius of 9 inches and a slant height of 6 inches _____

9. a radius of 2.5 meters and a slant height of 5 meters _____

10. a diameter of 16 feet and a slant height of 10 feet _____

11. a diameter of 8.5 centimeters and a slant height of 7 centimeters _____

12. a radius of 7 inches and a slant height of 9 inches _____

13. a radius of $6\frac{1}{4}$ feet and a slant height of 6 feet _____

Practice

12.7 Spheres

Find the surface area and volume of each sphere shown. Round answers to the nearest hundredth.

1.

$r = 13$ ft

_____

2.

$d = 5$ in.

_____

3.

$d = 3.5$ cm

_____

4.

$C = 18\pi$ m

_____

Find the surface area and volume of each sphere described. Round answers to the nearest hundredth.

5. a radius of 6 m   _____

6. a diameter of 15 in.   _____

7. a diameter of 20 ft   _____

8. a radius of 8.6 cm   _____

9. a radius of 13 cm   _____

10. a diameter of 7.6 m   _____

11. a diameter of 6.5 in. _____

12. a diameter of 12.75 m _____

13. a radius of $6\frac{1}{3}$ yd   _____

14. a radius of $7\frac{1}{4}$ ft   _____

15. a diameter of 6.25 m _____

16. a diameter of 8.75 cm _____